GW01227702

Practical Gemmology

By the same author

The Diamond Compendium

Practical Gemmology

DeeDee Cunningham

NAG Press
an imprint of Robert Hale · London

First published in Great Britain in 1955
This edition © 2011 DeeDee Cunningham
and the estate of Robert Webster

ISBN 978-0-7198-0431-1

NAG Press
Clerkenwell House
Clerkenwell Green
London EC1R 0HT

NAG Press is an imprint of Robert Hale Limited

www.halebooks.com

The right of DeeDee Cunningham and Robert Webster to be
identified as the authors of this work has been asserted by them
in accordance with the Copyright, Designs and
Patents Act 1988

A catalogue record for this book is available from the British Library

2 4 6 8 10 9 7 5 3 1

Typeset by e-type, Liverpool
Printed in the UK by the MPG Books Group Ltd,
Bodmin and King's Lynn

Contents

	Preface	7
Lesson 1	Introduction	9
Lesson 2	The Chemistry of Gemstones	15
Lesson 3	Crystallography	31
Lesson 4	Hardness, Cleavage, Parting and Fracture	47
Lesson 5	Density and Specific Gravity	59
Lesson 6	Light	73
Lesson 7	Refractive Index and Measurement of Refractive Index	97
Lesson 8	Polarized Light and Colour in Gem Distinction	121
Lesson 9	Luminescence, Electrical and Thermal Properties, X-rays and Magnetism	142
Lesson 10	Loupe and Microscope	159
Lesson 11	Primary Gem Species	170
Lesson 12	Synthetic Gemstones	205
Lesson 13	Simulants	234
Lesson 14	Composite Stones	243
Lesson 15	Gemstone Altering Treatments	251
Lesson 16	Gemstone Cutting and Styles of Cut	267
Lesson 17	The Pearl	283
Lesson 18	Organic Gems	302
Lesson 19	Secondary Gem Species	312
	Index	325

Preface

For six decades since its publication in 1943, Robert Webster's *Practical Gemmology* was standard reading for professional gemmology courses. It was the first gemmological book I ever purchased. And so it was a great pleasure to be entrusted with the task of rewriting it so that the title may be returned to its esteemed place among professional gemmological education books.

A great deal has changed in the field of gemmology since Webster's last edition in 1976. New diamond simulants have come to the marketplace, high-tech treatments and synthesis methods have been developed and knowledge of the theory and science of gemmology has progressed exponentially. The lesson format and the basic order, composition and illustrations of Webster's lessons have been retained, but the book has been completely rewritten to the current vernacular and the contents have been expanded, taking the book from being one of introductory interest only to one that will also serve the diploma curriculum. Each lesson is independent and information is presented serially so that any required level of understanding may be achieved.

More information is given about gemstone species. Webster originally wrote very little on the gem species, as he intended the book to be complemented by data from other books. I have covered the key properties of the curriculum gemstones as it enhances understanding of the individual lessons to have this information conveniently gathered together with theory in the same book. All of the organic gems, with the exception of pearl, which has its own chapter, are in one chapter, and inorganic gemstones are divided into Primary Gem Species, the principal stones of the professional curriculum, and Secondary Gem Species, those stones that are encountered less often.

Thank you to my publisher, Robert Hale Limited. I am indeed privileged to work with chairman, John Hale, publicity manager Paola

Motka, and my trusted editor Nikki Edwards. I thank the Gemological Institute of America for use of their reference library and their team of genius librarians, Paula Rucinski, Peggy Tsiamis and Rose Tozer, and special thanks go to my friends Cathy Jonathan and Gus Pritchett who always seem to find what I want, no matter how obscure, and who continue to assist me when I am at home working in Toronto. Richard Cartier FGA has generously reviewed every chapter and I thank him on behalf of myself and the students who will benefit from the incorporation of his suggestions.

I am aware that I have been 'heavy-handed' with this rewrite but to do any less would not respect the prior work. I am certain Robert Webster would rather I did the subject justice so that *Practical Gemmology* may continue to benefit those who study gemmology worldwide.

DeeDee Cunningham
Toronto, Canada

LESSON 1

Introduction

Gemmology is the art and science of evaluating the properties of and identifying gemstones based on a range of technical aspects. Although the beauty and lore of gems have been studied for hundreds of years, a scientific basis was first introduced at the turn of the twentieth century in response to the need to recognize cultured pearls. As the materials used for personal adornment are so varied, so are the true sciences to which students must turn to understand gem materials. Modern gemmology incorporates the non-destructive tests of chemistry, mineralogy, petrology and geology, as well as the scientific theories of physics, optics and crystallography, and the theory and use of specialized instruments. Products of nature can be either: (1) organic – formed by once-living organisms; these are from the animal and vegetable kingdoms; or (2) inorganic – formed by earth processes not involving organic life; these are from the mineral kingdom. Most gems are minerals; however, those derived from animals, such as pearl and ivory, and those of vegetable origin, such as amber and jet, involve zoology, biology and botany.

Gemmological knowledge is constantly evolving as high-tech processes and a rising number of synthetics and imitations enter the marketplace, in addition to newly discovered gemstones and advancements in science and research. In any study of gems it is usual, and indeed advantageous, to approach the subject from the standpoint of mineral gemstones, leaving those of animal and vegetable origin for a separate later discussion, a method that is followed in this book. We begin with an introduction to the planet Earth.

The Earth

The Earth is a sphere with a slight bulge around the equator, an ellipsoid, with an equatorial circumference of 40 075km and a diameter at the equator of 12 756km. It is the third planet out from the star we refer to as the Sun. Earth is the fifth largest of the eight planets in our solar system, but it is the largest of the terrestrial, meaning solid or non-gas, planets. With an equatorial radius of 6378km, the interior is layered into geologic components known as the crust, mantle and core.

The inner core has a radius of ~1200km and is solid, composed primarily of iron. The outer core extends 2250km beyond this to a radius of ~3450km and is liquid, composed of iron mixed with some nickel and trace amounts of other elements. Outside of this is the mantle, a highly viscous solid/plastic layer that is ~2850km thick. It is divided into the zones of: upper mantle 35–660km below the surface and lower mantle 660–2890km below the surface. The temperature of the mantle varies between 500°C and 900°C near the crust to 4000°C near the core. The crust, the part on which we live, floats above the mantle. It constitutes less than 1% of the Earth by volume and ranges from 5km to 70km deep. The thinnest parts are oceanic crust, 5–10km thick, under ocean basins, and the thickest parts are continental crust, averaging 30–50km in thickness. The main mineral constituents of oceanic crust are silica and magnesium and the main mineral constituents of continental crust are silica and alumina. Most gem minerals are formed in the crust; the exception is diamond, it forms in the upper mantle. The crust is composed of three basic kinds of rocks, each type determined by the process of formation.

Igneous, metamorphic and sedimentary rocks

Igneous rock, from the Latin *ignis* meaning 'fire', is formed by molten rock, termed magma, which has cooled and solidified. The molten rock is derived from partial melts of pre-existing mantle and/or crustal rocks. It forms below the surface as intrusive, plutonic or abyssal rock which cools slowly, and as a result is characteristically coarse-grained with the minerals often eye-visible or at the surface as volcanic or extrusive rock which is fine-grained as a result of quick cooling. A very coarse-grained type called pegmatite is an important source of gem materials. Granite and obsidian are examples of igneous rocks, the first formed by slow cooling and the second by quick cooling of molten rock. The crust is a mostly igneous product of past mantle melting in which incompatible elements separate out from mantle rock and the lighter material rises through cracks and fissures and cools at the surface. Igneous rocks account for 95% of the upper crust. But the evidence of this is hidden beneath a thin layer of sedimentary and metamorphic rocks at the surface.

Metamorphic rock is a product of the transformation of a pre-existing rock termed the protolith. In the process of metamorphism the protolith, which may be any rock type – igneous, sedimentary or pre-existing metamorphic rock – is subjected to high heat (>150°C) and high pressure (>1500 bars) causing physical and chemical change. The high pressure may be produced by the intrusion of igneous rock or be the product of earth movements. Marble is produced from limestone by such an action, and if the original limestone contained impurities these may also reform, producing other minerals in the marble.

Sedimentary rocks are laid down in layers, formed by deposition and consolidation of both mineral and organic material. Transporting agents such as wind, ice and water flows carry particles in suspension and then eventually deposit the particles after which the cumulative weight of the particles, the overburden pressure, compacts them into rock over thousands of years by a process known as lithification, which squeezes the sediment into layered solids. After deposition, rocks undergo physical, chemical and biological changes during and after lithification. Sedimentary rocks are also subject to surface weathering *in situ*. Sandstone and limestone are sedimentary rocks. Scientists have learned a great deal about the history of the Earth from studying sedimentary rock, as the differences between layers are indicative of past changes in the environment. They are also a valuable source of fossils as their formation conditions, unlike those of the other two rock types, do not require high temperature and pressure capable of destroying fossils. Sedimentary rocks cover most of the Earth's continents.

Table 1.1 Geologic layers of the Earth

Component layer	*Depth in km*
Crust	0–35*
Mantle	35–2890
Upper mantle	35–660
Lower mantle	660–2890
Core	2890–6378
Outer core	2890–5100
Inner core	5100–6378

* Average value – ranges 5–70 in the extreme

Study of the geological formation environments of gem minerals is complicated by the fact that a number of gems form in more than one type of environment and in some cases their formation conditions remain debatable. Often their true geologic source is concealed because the minerals have formed deeper down in one type of rock and are transported to the surface in volcanic eruptions of igneous magma. Also, many minerals form by the process of metasomatism, the changing of the composition of a rock by removal or introduction of chemical elements by hydrothermal and other fluids. Hydrothermal activity, in simple terms the percolation of hot fluids, occurs with both metamorphic and igneous rocks. Nephrite, amethyst, tanzanite, tsavorite garnet, emerald and topaz are among those gems found in hydrothermal veins.

Many mineral gems are associated with the coarse-grained plutonic igneous rock pegmatite, where the slower rate of cooling below the surface allows the atoms to assemble into orderly lattice structures building up a crystal which may, if enough time is allowed and conditions are favourable, grow to a considerable size. Such processes form zircon, ruby, sapphire, amethyst, topaz and spinel. Chemical change in metamorphic rocks by hot magma erupting into cooler rocks forms emerald and corundum. Metamorphic shearing and subsequent crushing of rocks produce garnet, jadeite and andalusite. Sedimentary rocks are the source of few primary gem materials with the exception of opal, formed by the interaction of sedimentary pore-fluids and groundwater, turquoise and organics such as amber and jet.

Rocks, minerals and gemstones

A mineral is a naturally occurring inorganic, usually crystalline, homogenous substance with a definite atomic structure, a specific chemical composition and physical properties that are practically the same for all specimens. Minerals may be pure elements, such as gold, platinum, or carbon in the case of diamond, or they may be very complex compounds. A rock is a naturally occurring solid aggregate of two or more minerals which may vary considerably in relative proportion and which may be easily separated. Common granite is a rock composed of three easily observed minerals: feldspar, mica and quartz; each of these three minerals is a definite chemical compound and may be mixed in almost any proportion to form granite. Inorganic gemstones are almost always one mineral. A notable exception is the rock lapis lazuli composed of different coloured minerals. The other exceptions to minerals as gemstones are of course those from the animal and vegetable kingdoms. There are over 4000 known minerals, but only about 100 of these have the qualities necessary to be prized as gems and, of these, about 20 constitute the major gemstones of the curriculum.

The first of the cardinal virtues of a gemstone is undoubtedly beauty – through transparency and depth of colour in the ruby and emerald, through colour only in opaque gems such as turquoise, through fire, brilliance and the absence of colour in diamond, and through various phenomena such as 'play of colour' in opal. It is only rarely that a rough gem or crystal is mounted in jewellery; most often the lapidary in cutting the stone brings out its latent beauty.

As valuable ornaments, gemstones should be able to resist abrasive influences that tend to destroy beauty and lustre, making durability the second of the cardinal virtues. These influences include sand and dust particles in the air, being rubbed against other objects, and chemical damage caused by contact with everyday substances and pollutants in the atmosphere. Durability is primarily governed by hardness and toughness and, in general, gemstones are hard, tough minerals. Glass imitations, referred to often as *paste*, are not durable; they show signs of wear even after a short time because they are softer than the sand and dust particles in the air, which are quartz with a Mohs hardness of 7. They are also readily attacked by the chemical action of the sulphur in the atmosphere. Chemical stability and the stability of colour upon exposure to light are also considerations for the virtue of durability.

Often of far greater influence than beauty or durability is the third virtue, rarity. A mineral may be fairly common, yet really fine pieces suitable for cutting may be quite rare. An example is emerald; a near-flawless emerald of fine colour is exceedingly rare and may command a higher price than diamond. The law of supply and demand, often influenced by fashion, governs to a certain extent the rarity of gemstones. Gems popular in fashion are often in short supply, creating a market rarity, and their price may rise rapidly. All gems are rare among minerals in general but rarity is relative; citrine is one of the more common gems but it is very popular and extremely attractive in fine colour and quality.

For classification, some minerals are divided into groups with similar characteristics and then further into species and varieties. Two important groups studied are garnet and feldspar. More often gems are categorized only into species with their own chemical compositions and characteristics – for example, corundum – and varieties of species based on colour or general appearance – for example, yellow sapphire and star ruby.

Gems can be found in: (1) primary deposits – the place of their formation or, in the case of diamond, its transportation to the surface; (2) alluvial deposits – secondary deposits to which the gems have been carried by the action of elements such as wind, rivers, rain and glaciers; and (3) eluvial deposits – residual deposits formed when weathering released the gems and left them in place without transporting them away from parent rocks.

Historically, gemstones have been divided into two classes as precious and semi-precious stones. Precious stones, according to the ancient Greeks, were diamond, ruby, emerald and sapphire, and over time pearl and black opal often came to be included – all stones for which the value of fine specimens is high and demand is fairly constant. Such a traditional division is quite arbitrary and does not accurately reflect the modern-day trade. Some stones are generally inexpensive, as is garnet, but there are green varieties of garnet, namely tsavorite and demantoid, that can be more expensive than diamond. The term 'semi-precious' is obsolete. Any attractive in-demand mineral or organic material with the requisite qualities of a gemstone is a gemstone, full stop.

LESSON 2

The Chemistry of Gemstones

All material things we know, whether solids, liquids or gases, are composed of one or more fundamental substances known as *elements*, of which *atoms* are the basic building blocks. Most materials are compounds consisting of two or more elements in fixed proportions that can be analysed by chemical means into simpler substances. Elements are composed of the same types of atoms and are not reducible to simpler substances by chemical means. In our environment, naturally occurring stable elements include solids such as iron and sulphur, liquids such as mercury and bromine, and gases such as oxygen and nitrogen.

The science of chemistry is the study of the composition, structure, behaviour and properties of matter and the laws that govern changes in matter by chemical reaction at the molecular level. Chemical reactions always involve a change to the nature of the substances involved. It is important to understand the difference between a mixture of two substances and a chemical compound formed between two substances. Iron filings and sulphur can form a mixture in which each substance is present in any proportion and the two substances can be readily separated: the iron can be removed by means of a magnet. When iron and sulphur are combined together in definite proportions to produce the chemical compound iron sulfide, this compound has properties very different from those of either iron or sulphur and it is no longer possible to separate the two by physical means. This distinguishes chemical properties from physical properties. Magnetism is a physical property of iron that can be tested or observed without permanently changing the iron; whereas the fact that iron can be dissolved in nitric acid is a chemical property which, when tested or observed, involves changes to both substances. In gemmology, physical properties are more important than chemical properties because they can be observed or measured without altering stones. Chemical tests, however delicate, always involve changing what is tested, and for a gemstone any change would be destructive. None the less, the properties of a gem depend upon both: (1) the chemical elements present; and (2) the interaction of those elements within the crystal structure. Crystallography is discussed in the following chapter; here is a presentation of basic chemistry as it applies to gemmology.

Atoms

Even in Roman times, the poet and philosopher Lucretius surmised that all matter could be explained in terms of the motions and interactions of tiny atoms in empty space. Early in the nineteenth century John Dalton postulated that elements are made of tiny particles called atoms; that all atoms of an element are identical in nature but different from atoms of other elements; and that atoms cannot be created, divided or destroyed when elements chemically combine to form a compound or are chemically removed from a compound; all that ever changes is the way the atoms are grouped together.

An atom is the basic unit of matter and the smallest amount of a chemical element that can exist alone and retain the properties of the element. Atoms cannot be chemically subdivided; however, they are not indivisible as once was thought by early philosophers. Sub-atomic particles make up the atom. In the centre of the atom in the nucleus there are positively charged *protons* and electrically neutral *neutrons*. These particles weigh the same. Orbiting around the nucleus are negatively charged *electrons* that weigh virtually nothing, only about 1/1850th of a proton or neutron. The basic structure is a dense nucleus surrounded by an almost weightless cloud of electrons with random location within specific orbitals, sometimes called *shells*. Within the atom, like charges repel and opposite charges attract. Electrons are held in their orbits by electromagnetic force in which the positive protons pull the negative electrons towards the nucleus. Within the nucleus, the stronger nuclear forces prevent the positive charges of the protons from repelling one another and splitting the nucleus apart. A carbon atom has six protons, six neutrons and six electrons. It is a stable atom with an equal number of protons and neutrons and is electrically neutral.

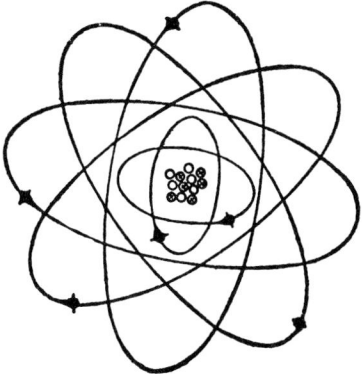

Figure 2.1 Schematic diagram of the structure of a carbon atom: in the nucleus the six circles with a cross represent protons and the six empty circles represent neutrons. Surrounding the nucleus the six black discs with cross bars represent electrons in their orbits.

Chemical elements

Minerals in general are either: (1) elements such as carbon (diamond or graphite), gold, and platinum; or (2) compounds consisting of two or more elements in fixed proportions such as aluminium oxide Al_2O_3 (corundum) or silicon dioxide SiO_2 (opal or quartz). In order to understand the chemical composition of gems one must understand the basic nature of chemical elements. Elements are pure substances consisting of identical atoms, defined by the atomic number of one atom of which they are comprised. The atomic number corresponds to the number of protons in the atom, which is constant for that kind of atom. Carbon has six protons and is element number six.

An element also has a distinctive atomic mass number reflecting the relative weight or, more properly, the relative atomic mass, of its atoms. Atomic mass is the number of protons plus the number of neutrons. This is a bit complex because not all atoms of an element have the same number of neutrons. The variations are termed *isotopes*: atoms of the same chemical element with a different atomic mass. Hydrogen typically has one proton and one electron. An isotope of hydrogen, known as deuterium – or heavy hydrogen – has one neutron. Deuterium is then twice as heavy as hydrogen because of the neutron, but they are still both the element hydrogen as defined by the one proton. The number of protons determines an element's electric charge, which in turn determines the number of electrons in an electrically neutral atom. This is the reason that atomic number, rather than atomic mass, is the defining characteristic of an element. Both hydrogen and deuterium have one electron, and therefore react the same chemically.

Relative atomic mass is the average of the atomic masses of all of that element's isotopes, weighted by abundance, relative to the atomic mass unit (u). Due to the averaging process this is usually a number with many decimal places. Oxygen always has eight protons but it can have any of eight, nine or ten neutrons as there are three stable isotopes for oxygen. This gives mass numbers of 16, 17 and 18 respectively and they are known as oxygen-16, oxygen-17 and oxygen-18 and by the symbols ^{16}O, ^{17}O and ^{18}O. Oxygen as we know it is 99.76% ^{16}O with the remainder made up of the stable isotopes mentioned, as well as ten other unstable isotopes ranging in mass from 12 to 24. This gives oxygen a relative atomic mass of 15.9994. Atomic mass was originally calculated against that of the lightest atom, hydrogen, taken as unity and then for a time against oxygen taken as 16. It has now been standardized and is taken to be the ratio of the average mass per atom to $^{1}/_{12}$ of the mass of an atom of carbon (u), specifically ^{12}C as this is fixed by definition at exactly 12.

Elemental chemical knowledge was made orderly when Russian chemist Dmitri Mendeleev discerned a periodic similarity among the elements when they were arranged in order of their atomic mass. His tabulation of the known elements into groups according to his 'periodic system' brought out recurring trends in the properties of elements such as: the group 1 alkali metals – lithium, sodium, potassium, etc.; the group 2 alkaline earth metals – beryllium, magnesium, calcium, etc.; and the group 17 halogen elements – fluorine, chlorine, bromine, iodine, etc. In 1868, helium was discovered and it marked a special group on the periodic table: that of noble gases. Noble gases rarely participate in chemical reactions and are said to be chemically inert. Of these gases, argon is the most plentiful.

In actuality, the first listing of elements was put forth by Antoine Lavoisier in 1789 and others before Mendeleev attempted classification systems, but Mendeleev's true accomplishment was that he was able to use the gaps in his system to foretell the existence of new elements

– some of which were discovered in his lifetime: gallium in 1875 and germanium in 1886 – having properties consistent with their predicted places in the table. At present, there are 117 known chemical elements, 94 of which occur naturally on Earth and 80 of which have stable isotopes. Each of the chemical elements is represented by an accepted symbol. Table 2.1 lists the first 94 elements along with their atomic numbers, atomic masses, symbols and valences.

Table 2.1 Chemical Elements: symbols, weights and valence

Atomic number	Element	Symbol	Atomic weight	Valence
1	Hydrogen	H	1.00794	1
2	Helium	He	4.002602	0
3	Lithium	Li	6.941	1
4	Beryllium	Be	9.012182	2
5	Boron	B	10.811	3
6	Carbon	C	12.0107	4
7	Nitrogen	N	14.0067	3
8	Oxygen	O	15.9994	2
9	Fluorine	F	18.9984032	1
10	Neon	Ne	20.1797	0
11	Sodium	Na	22.98977	1
12	Magnesium	Mg	24.305	2
13	Aluminium	Al	26.981538	3
14	Silicon	Si	28.0855	4
15	Phosphorus	P	30.973761	5
16	Sulphur	S	32.065	6
17	Chlorine	Cl	35.453	5
18	Argon	Ar	39.948	0
19	Potassium	K	39.0983	1
20	Calcium	Ca	40.078	2
21	Scandium	Sc	44.95591	3
22	Titanium	Ti	47.867	4
23	Vanadium	V	50.9415	5
24	Chromium	Cr	51.9961	6
25	Manganese	Mn	54.938049	4

Atomic number	Element	Symbol	Atomic weight	Valence
26	Iron	Fe	55.845	3
27	Cobalt	Co	58.9332	4
28	Nickel	Ni	58.6934	2
29	Copper	Cu	63.546	2
30	Zinc	Zn	65.409	2
31	Gallium	Ga	69.723	3
32	Germanium	Ge	72.64	4
33	Arsenic	As	74.9216	5
34	Selenium	Se	78.96	6
35	Bromine	Br	79.904	7
36	Krypton	Kr	83.798	2
37	Rubidium	Rb	85.4678	1
38	Strontium	Sr	87.62	2
39	Yttrium	Y	88.90585	3
40	Zirconium	Zr	91.224	4
41	Niobium	Nb	92.90638	5
42	Molybdenum	Mo	95.94	6
43	Technetium	Tc	98*	7
44	Ruthenium	Ru	101.07	6
45	Rhodium	Rh	102.9055	6
46	Palladium	Pd	106.42	4
47	Silver	Ag	107.8682	2
48	Cadmium	Cd	112.411	2
49	Indium	In	114.818	3
50	Tin	Sn	118.71	4
51	Antimony	Sb	121.76	5
52	Tellurium	Te	127.6	6
53	Iodine	I	126.90447	7
54	Xenon	Xe	131.293	6
55	Caesium	Cs	132.90545	1
56	Barium	Ba	137.327	2
57	Lanthanum	La	138.9055	3
58	Cerium	Ce	140.116	4
59	Praseodymium	Pr	140.90765	4
60	Neodymium	Nd	144.24	3

THE CHEMISTRY OF GEMSTONES

Atomic number	Element	Symbol	Atomic weight	Valence
61	Promethium	Pm	145*	3
62	Samarium	Sm	150.36	3
63	Europium	Eu	151.96	3
64	Gadolinium	Gd	157.25	3
65	Terbium	Tb	158.92534	3
66	Dysprosium	Dy	162.5	3
67	Holmium	Ho	164.93032	3
68	Erbium	Er	167.259	3
69	Thulium	Tm	168.93421	3
70	Ytterbium	Yb	173.04	3
71	Lutetium	Lu	174.967	3
72	Hafnium	Hf	178.49	4
73	Tantalum	Ta	180.9479	5
74	Tungsten	W	183.84	6
75	Rhenium	Re	186.207	7
76	Osmium	Os	190.23	6
77	Iridium	Ir	192.217	6
78	Platinum	Pt	195.078	6
79	Gold	Au	196.96655	5
80	Mercury	Hg	200.59	2
81	Thallium	Tl	204.3833	3
82	Lead	Pb	207.2	4
83	Bismuth	Bi	208.98038	5
84	Polonium	Po	209*	6
85	Astatine	At	210*	7
86	Radon	Rn	222*	6
87	Francium	Fr	223*	3
88	Radium	Ra	226*	2
89	Actinium	Ac	227*	3
90	Thorium	Th	232.0381	4
91	Protactinium	Pa	231.03588	5
92	Uranium	U	238.02891	6
93	Neptunium	Np	237*	6
94	Plutonium	Pu	244*	6

* For longest lived isotope

Valence, ions and bonding

Atoms join together to form chemical compounds in a process called bonding. Protons and neutrons are heavy and held together by strong atomic forces so it is the lighter, more mobile, electrons that play the important role in bonding. Electrons orbit the nucleus in concentric shells called energy levels. These can be thought of as similar to the orbits in which planets circle the sun – except that unperturbed planetary orbits are actually two-dimensional in a plane fixed in space, whereas electron orbits are three-dimensional shapes like the skin of a balloon, and the electrons reside within these shells travelling at the speed of light, so we can never know the location of any electron within a shell. Strict rules govern the number of electrons required to fill a shell as well as the distance that shell resides from the nucleus. They are arranged in a manner that maintains the lowest energy level when full. Shell number 1, the innermost shell, takes a maximum of two electrons. For the first two elements this shell is the outer shell. Helium, the element with two electrons, has a full shell. The elements with more than two electrons have electrons arranged in shells 1–7 with the outermost shell, whichever number it is, termed the valence shell, always taking a maximum of eight electrons. Carbon has six electrons: two in the first shell and four in the second shell. Neon has ten electrons: two in the first shell and eight that fill the second shell. The full outer shell makes helium and neon chemically inert and therefore noble gases. In bonding, atoms seek stability and strive for a full outer shell like that of noble gases. This can be achieved by losing, gaining or sharing electrons. Valence is the combining power of an atom of an element as determined by the number of electrons it will lose, gain or share when reacting with other atoms; this is dictated by the number of electrons in the outer shell, termed valence electrons.

Atoms with valences of 1, 2 or 3 having one, two or three electrons in the outer shell tend to lose those electrons in interactions with atoms having seven, six or five electrons. Similarly, atoms with valences of 5, 6 or 7 having five, six or seven electrons in the outer shell tend to gain electrons in interactions with atoms having three, two or one electrons. Atoms with four electrons usually share electrons. Some confusion can arise with the two different ways valence is expressed. Atoms with seven electrons have a valence of 7; however, this is sometimes expressed as −1, reflecting the fact that the element would accept one electron,

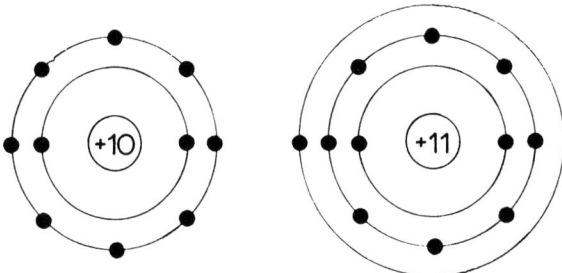

Figure 2.2 Bohr-Rutherford-style diagrams depicting the structure of neon and sodium atoms. Neon has 10 protons as indicated in the centre and 10 electrons: two in the inner shell and eight in the outer shell. This fills the outer shell and establishes neon as a chemically inert noble gas. Sodium has 11 protons as indicated in the centre and 11 electrons: two in the inner shell, eight in the second shell, and one in the outer shell giving it a valence of 1. The next atom similar to sodium is potassium with 19 protons and 19 electrons. It has the third electron shell completely full with eight electrons and then has a shell farther out with one electron. This also gives it a valence of 1 and establishes the two elements as having similar chemical characteristics.

resulting in a negative charge, to achieve a stable shell. By this system atoms have valences of −1, −2, −3 when electrons are accepted to yield a negative charge, and +1, +2 and +3 when electrons are given up to yield a positive charge, as well as 4, which may be considered either positive or negative. When atoms gain or lose electrons they become charged particles called *ions*. Ions are the smallest particles of matter that take part in chemical reactions. They are incapable of existing alone and readily take part in reactions to form molecules: stable, electrically neutral groups of atoms. A gain of electron(s) produces a negatively charged ion called an *anion* and a loss of electron(s) produces a positively charged ion called a *cation*.

The hydrogen atom having only one proton and one electron will lose the electron producing a positively charged cation consisting of one single proton. The removal of the electron from the hydrogen isotope deuterium produces a cation consisting of one proton and one neutron, known as a *deuteron*. And, removal of the two electrons from the helium atom produces a cation consisting of two protons and two neutrons known as an *alpha particle*. These particles were used in the artificial coloration of diamond by irradiation.

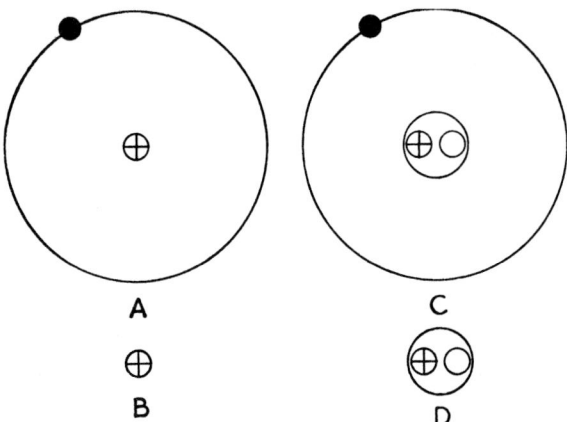

Figure 2.3
A Stylized diagram of a hydrogen atom with one proton in the nucleus and one surrounding electron.
B Loss of the electron leaves a positively charged particle, a cation, consisting of one proton.
C Stylized diagram of the hydrogen isotope deuterium with one proton and the extra neutron in the nucleus. One electron surrounds.
D Loss of the electron leaves a positively charged particle, a cation, consisting of one proton and one neutron, known as a deuteron.

The two principal types of bonds are ionic and covalent. Ionic bonds are formed when electrons are transferred by loss or gain and covalent bonds are formed when electrons are shared. Since there is no transfer of electrons with covalent bonding, ions are not produced. An example of an ionic bond where electrons are given off and taken in to achieve a stable outer shell occurs with the formation of table salt:

$$Na^+ + Cl^- \rightarrow NaCl$$

Sodium, valence 1, has one electron in the outer shell that it loses to form a positively charged cation. Chlorine, valence 7, is one electron short of a full shell and it gains this electron from sodium creating a negatively charged anion. The ionic bonding of the anion and the cation produces the stable, electrically neutral salt. Ionic bonding is present in many gems including corundum and spinel.

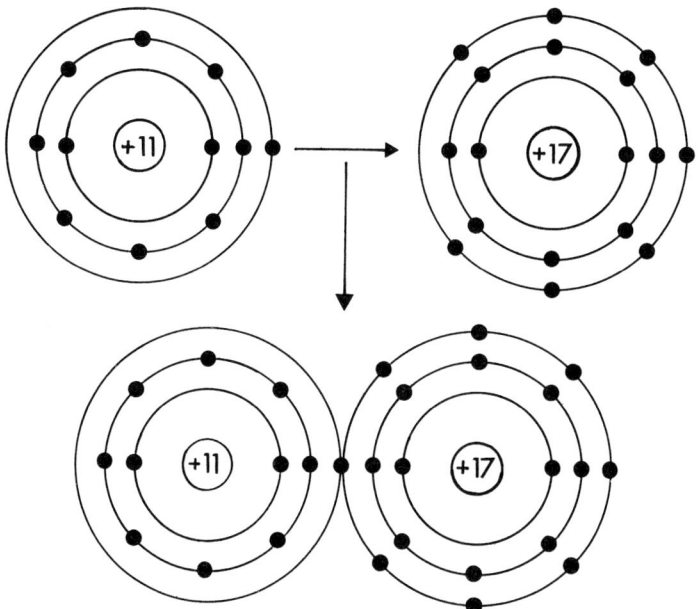

Figure 2.4 Ionic bonding of sodium and chlorine atoms forms sodium chloride. There is one electron in the outer shell of the sodium atom on the left. This electron is given off to leave a complete outer shell and creates a positively charged ion called a cation. There are seven electrons, one less than a full outer shell, in the chlorine atom on the right. Chlorine gains the electron given up by the sodium to create a complete outer shell, and the negatively charged ion is called an anion. Combining the opposite charges of the cation and the anion produces ionic bonding and forms the molecule of sodium chloride. This compound, with the same total count of sub-atomic particles as the two separate atoms, is stable with a full outer shell like that of a noble gas.

Diamond bonds covalently as carbon has four electrons in the outer shell and in the diamond structure each carbon atom forms four covalent bonds with four other carbon atoms to produce the effect of a full outer shell in all atoms. It is common for gems to contain combination bonds involving both ionic and covalent bonds and the Si–O bonds of silicate crystal structures like quartz are about half covalent and half ionic.

Molecules, chemical compounds and formulas

Compounds are pure chemical substances consisting of two or more different chemical elements in fixed proportions that can be separated into simpler substances by chemical reactions. Most gemstones are compounds but diamond is a single element: carbon. Molecules are stable, neutral groups of at least two atoms held together by strong covalent bonds. Molecules can bond together with atoms or other molecules to form compounds. Some elements, such as hydrogen and oxygen, become more stable when they form molecules; the atoms join in pairs as H_2 and O_2 and as diatomic molecules they join in compounds. Elements that are more stable in the single state join in compounds as single atoms.

Chemical compounds may be quite simple, such as titanium dioxide, TiO_2, which when in crystal form is the mineral rutile, or more complicated such as that of beryl, $Be_3Al_2(SiO_3)_6$. In titanium dioxide, TiO_2, titanium has a valence of four and oxygen has a valence of two. Therefore, it takes two oxygens to balance the titanium and produce titanium dioxide.

A chemical reaction to produce a compound can always be expressed as an equation utilizing the established chemical symbols. An equation illustrating the formation of corundum is:

$$4Al + 3O_2 \rightarrow 2Al_2O_3$$

The number in front of the chemical symbol represents the number of atoms or molecules. The number in subscript after the chemical symbol represents the number of atoms in each molecule. Equations are balanced with an equal number of atoms on each side of the equation – in this case, four aluminium and six oxygen. Aluminium has three electrons in the outer shell and oxygen has six. In order to bond and produce a stable compound with eight electrons in the outer shell the oxygen needs two electrons. Aluminium has three to lose, therefore to satisfy this condition it requires involvement of six (2 x 3) electrons; this is achieved by combining two aluminium atoms having a total of six electrons to lose and three oxygen atoms having a total of six electrons to gain. But, since oxygen is diatomic and participates in the compound as a 2-atom molecule the amount of aluminium must be doubled to four atoms to balance the equation. Expressed in words, four aluminium atoms combine with three molecules of oxygen to produce two molecules of aluminium oxide. A molecule of aluminium oxide is the smallest part of the compound that can exist alone, but it is not the smallest part of a crystal. That is the unit cell, which is discussed in the following chapter.

Most gemstones are oxides, carbonates, phosphates or silicates. Oxides consist of a metal combined with oxygen, which produces minerals that are generally hard and resistant to chemical attacks, such as Al_2O_3, the mineral corundum. Carbonates are formed by the action of carbonic acid on metals, which produces minerals that are characteristically soft and easily attacked by acids, such as $CaCO_3$, the mineral calcite. Phosphates are formed by the action of phosphoric acid on metals, which produces minerals that are soft and readily attacked by acid, such as $Ca(F,Cl)Ca_4(PO_4)_3$, the mineral apatite. Silicates are derived from the action of silicic acid on metals, which produces gems that are hard and durable because of the strong covalent bonds, such as $ZrSiO_4$, the mineral zircon.

Silicates are the most prevalent gem group as oxygen and silicon are the two most abundant elements in the Earth's crust. These two elements combine with the other elements such as aluminium, iron, magnesium, etc., to form silicate minerals. Quartz, SiO_2, is found worldwide and constitutes approximately 12% of the upper mantle region and the Earth's crust. Silicates account for about 60% of gem minerals.

Transition elements

Some elements such as iron and copper have more than one valence. This is due to complications in which valence electrons from two shells participate in bonding under certain conditions. Iron can have a valence of either 2 or 3. When it has a valence of 2, termed divalent, it forms ferrous compounds such as FeO, ferrous oxide; and when it has a valence of 3, termed trivalent, it forms ferric compounds such as Fe_2O_3, ferric oxide. Similarly, copper can have a valence of either 1 or 2. When it has a valence of 1, termed monovalent, it forms cuprous compounds such as Cu_2O, cuprous oxide; and when it is divalent it forms cupric compounds such as $CuSO_4$, cupric sulfate.

Transition elements, also known as transition metals, occupy the long middle portions of the periodic table. Their first 18 electrons fill shells in standard order: two electrons in shell 1, eight electrons in shell 2 and eight electrons in shell 3. After this, electrons 19 and 20 go into shell 4 and electrons 21, 22 and 23 go into shell 3. The energy levels of electrons 21, 22 and 23 are lower than those of electrons 19 and 20 and this causes them to fall back into shell 3 as opposed to continuing to fill shell 4. Table 2.2 illustrates the electron configurations for the first 36 elements. Elements 22–29 – titanium, vanadium, chromium, manganese, iron, cobalt, nickel and copper – are important in gemmology. They are characterized as being held together by relatively weak atomic forces and their particles can be stimulated into vibration by various forms of energy. When the form of energy is light, the vibration of the particles results in selective light absorption producing a perception of colour in the object. These elements are responsible for colouring many of the gemstones we know: chromium colours emerald and ruby, and sapphire is coloured by iron and titanium.

Isomorphous replacement

In the old days chemists toiled to make some sort of neat chemical formula from chemically complex minerals. Now X-ray crystallographers are able to reveal the essentially simple structure of chemically complex minerals, showing that the apparent chemical complexity is due to isomorphous replacement in the structure. Isomorphous replacement occurs when one ion or radical occupies the position of another ion or radical in the crystal structure. Radicals are complex ions such as hydroxyl $(OH)^-$, carbonate $(CO_3)^{-2}$ and phosphate $(PO_4)^{-3}$ that behave as single ions and can have valences of 1, 2 or 3 as indicated in the three examples. Radicals are strongly bound together. They are always anions and unstable by themselves, readily seeking to combine with other elements and compounds. When a gem is crystallizing, conditions may be such that two ions or radicals of the same size and electrical charge are present in the environment and so either one can fit into the crystal and fill the space without disrupting the structure. It is important to understand that this is not literal replacement. It is simply that the alternate ion or radical was there by pure chance and came to be incorporated into the crystal. The crystal formed from the start with this formula. For isomorphous replacement to be a complete series the ions or radicals have to be both the same charge and practically the same size.

Table 2.2 Chemical elements: electron shell configuration and groups

Atomic number	Element	Electron shell configuration	Group
1	Hydrogen	1	1
2*	Helium	2	18
3	Lithium	2,1	1
4	Beryllium	2,2	2
5	Boron	2,3	13
6	Carbon	2,4	14
7	Nitrogen	2,5	15
8	Oxygen	2,6	16
9	Fluorine	2,7	17
10*	Neon	2,8	18
11	Sodium	2,8,1	1
12	Magnesium	2,8,2	2
13	Aluminium	2,8,3	13
14	Silicon	2,8,4	14
15	Phosphorus	2,8,5	15
16	Sulphur	2,8,6	16
17	Chlorine	2,8,7	17
18*	Argon	2,8,8	18
19	Potassium	2,8,8,1	1
20	Calcium	2,8,8,2	2
21	Scandium	2,8,9,2	3
22	Titanium	2,8,10,2	4
23	Vanadium	2,8,11,2	5
24	Chromium	2,8,13,1	6
25	Manganese	2,8,13,2	7
26	Iron	2,8,14,2	8
27	Cobalt	2,8,15,2	9
28	Nickel	2,8,16,2	10
29	Copper	2,8,18,1	11
30	Zinc	2,8,18,2	12
31	Gallium	2,8,18,3	13
32	Germanium	2,8,18,4	14
33	Arsenic	2,8,18,5	15
34	Selenium	2,8,18,6	16
35	Bromine	2,8,18,7	17
36*	Krypton	2,8,18,8	18

* Noble Gas

They will not be exactly the same size, as no two are alike, but if they are within a narrow tolerance it leaves a little distortion of the space but crystallization continues. If the ions or radicals are the same size but differ in charge, then another ion or radical with the complementary charge will have to be incorporated to retain neutrality in the structure.

Several gem species show isomorphous replacement including garnets and feldspars. Peridot is in the middle of an isomorphous series but the end members are not attractive enough to be gems. Perhaps the best illustrative example is that of topaz because it involves an ion and a radical, both end members are still called topaz, and it produces unique effects on major diagnostic factors. Topaz is aluminium silicate with some hydroxyl and/or fluorine. The formula is $Al_2(F,OH)_2SiO_4$ where the fluorine ion (F) or the hydroxyl radical (OH) can occupy the same site and may be present in various amounts. The end members, $Al_2F_2SiO_4$ and $Al_2(OH)_2SiO_4$, are theoretical. The fluorine-rich topazes are colourless and blue with RI (refractive index) approximately 1.61–1.62, BI (birefringence) 0.010 and SG (specific gravity) averaging 3.56. The hydroxyl-rich topazes are yellow and brown with RI approximately 1.63–1.64, BI 0.008, and SG averaging 3.53. Usually, if RI increases then SG increases; however, in the case of topaz, the (OH) increases the RI but causes a lower SG, whereas the F lowers the RI but causes a higher SG. The RI and SG values of any specimen will vary accordingly depending on the proportions of F and (OH) present. The BI is higher for the fluorine-rich stones, but varies downwards as (OH) levels increase.

LESSON 3

Crystallography

Crystallography is the scientific study of the arrangement of atoms in crystals involving the classification and measurement of crystals. In 1669 Nicholas Steno observed that even though all quartz crystals did not look alike there was correspondence between the angles of similar faces and this correspondence was absolute regardless of the shape or size of the crystal. Crystallography established that the nature of any crystal is a three-dimensional latticework of atoms and that the properties of gemstones depend more upon the arrangement of atoms in that lattice than upon their chemical nature. Students often have difficulty appreciating the value of crystallography. Let it be said at the outset that knowledge of elementary crystallography lays the foundation for much of the later work. It allows us to understand the nature of gemstones and their properties for testing and identification, as well as the appearance of rough crystals. The study of crystallography is of value to all who handle gems. For example, lapidaries determine how to orient a stone for cutting based on understanding how the direction light travels through the stone affects its resulting colour; diamond cutters can avoid trying to cut a stone in the hardest direction or to polish in a cleavage direction; and stone setters can know which stones may break due to cleavage.

Crystals

All solid matter is either: (1) crystalline – with the atoms arranged in an orderly and symmetrical lattice structure; or (2) amorphous or non-crystalline – with atoms in random locations without structure. Most specimens that gemmologists encounter are crystalline; however, glass, plastics and a limited number of gems such as opal and amber are amorphous. A *crystal* is defined as a solid substance of definite polyhedral shape (meaning bounded by plane faces) with a regularly repeating three-dimensional lattice structure due to the action of inter-atomic forces on the established chemical composition during the change of matter from a liquid or gas to a solid state under suitable conditions. This can be remembered by the characteristics of a crystal: (1) an orderly and symmetrical lattice structure; (2) a definite external geometric shape bounded by more or less plane faces; and (3) physical and optical properties that may vary by direction from very slight to pronounced measure. These properties exist even in distorted crystals, broken fragments, or faceted gems with no sign of the original shape because they depend on atomic structure, not external shape. It is only that with well-formed crystals the external shape is present as a direct outward manifestation of internal orderly and symmetrical atomic structure.

Minerals that possess an orderly and symmetrical crystal lattice structure and directional properties but lack a geometrical external shape with plane faces are termed *massive crystalline*. Massive rose quartz lacks the external shape, but still possesses all the physical properties of crystalline quartz. *Cryptocrystalline*, a term meaning 'hidden crystalline', is the name given to a structure having a vast number of very tiny crystals in an unbroken mass, like that of chalcedony.

Some minerals can lose or be in the process of losing their original crystal structure and become amorphous, literally meaning without morphology, which may be interpreted as 'without structural shape'. Such minerals are termed *metamict* and their degeneration is due to lattice damage caused by radioactive elements either within the mineral itself or in its surroundings. Zircon is typically metamict because it has in its composition trace amounts of the elements thorium and uranium. The process takes a very long time and the degree of crystalline breakdown depends on the duration of exposure. The reduction in structure reduces both the RI and the SG.

When the same chemical composition assumes a different crystallographic structure it produces a completely different mineral with different physical and optical properties. Such minerals are called *polymorphs*. Diamond and graphite are polymorphs because both are atomically carbon but crystallographic differences give rise to their markedly different properties: diamond is the hardest natural substance on Earth and is exceptionally transparent, whereas graphite is very soft and opaque.

Diamond has carbon atoms bonded in a tetrahedral structure with all bonds strongly held by covalent forces. Graphite has carbon atoms arranged in layers with strong bonds within the layers and weak bonds between the layers; this allows the layers to slip away from one another easily, producing the quality that makes graphite ideal for pencil leads.

Every mineral has a distinctive sub-microscopic size unit referred to as the *unit cell*. It consists of the fewest number of atoms necessary to maintain all the properties of the crystal and is best thought of as a building block that is stacked and oriented the same way repeatedly billions and billions of times to form the crystal. Unit cells are the crystallographic equivalent of the molecule in chemistry. The unit cell of diamond is a face-centred cube.

Symmetry

Each mineral has a characteristic appearance with an established number, size, shape and position of crystal faces. Symmetry is defined as the regularity of shape and arrangement of crystal faces and the angles between them. Crystals are classified into 32 classes according to their symmetry, and these classes are grouped into seven crystal systems that have similar optical properties. It is important to understand these basics because the system a gemstone belongs to dictates its physical and optical properties, allowing the gemmologist to predict how it will behave under testing.

Elements of symmetry

Elements of symmetry are a way to describe the orderliness of crystal shapes. Each element of symmetry sets out a particular way to observe the relationship between identical faces of a crystal. The three most basic elements of concern are: plane of symmetry, centre of symmetry and axis of symmetry.

A *plane of symmetry (mirror plane)* is an imaginary flat surface in a perfectly developed crystal that will divide the crystal into two parts such that each half is the mirror image of the other: every edge, face and point on one side is directly opposite a similar edge, face and point on the other side. If a perfect crystal were cut in half along a plane of symmetry and placed on a front-silvered mirror, the reflection would replace

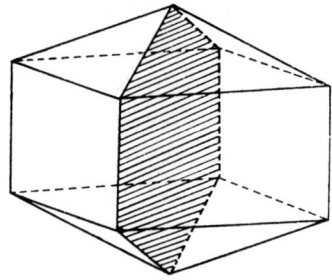

Figure 3.1 A plane of symmetry in a perfectly developed crystal. Each half is the mirror image of the other half.

the missing half. Hold your right and left hands with the palms away from you and the tips of the thumbs touching; a plane of symmetry runs vertically between the tips of the thumbs, as the right hand is the mirror image of the left.

A *centre of symmetry* is present when every point on a face of a perfectly developed crystal is diametrically opposite an identical point on a similar face on the opposite side at the same distance from the centre. In crystals with a centre of symmetry, any point on any surface in any direction from the centre will have a diametrically opposite duplicate. The number '96' written this way has a centre of symmetry but no other symmetry elements.

An *axis of symmetry* is an imaginary line extending into and through the centre of a perfectly developed crystal about which rotation will cause the crystal to show the same appearance two, three, four or six times in one complete revolution. The axes are termed diad or 2-fold,

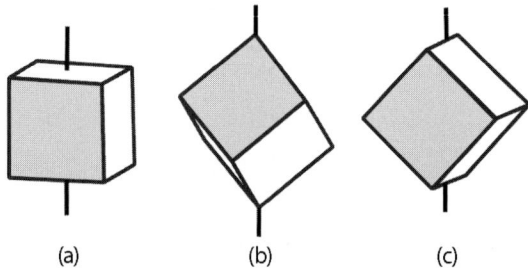

Figure 3.2 The axes of symmetry in a cube: (a) 4-fold; (b) 3-fold; and (c) 2-fold.

triad or 3-fold, tetrad or 4-fold and hexad or 6-fold. Crystals can have several axes of symmetry; there are 13 in total in a cube: three 4-fold, four 3-fold and six 2-fold. Cubes belong to the system with the highest level of symmetry: the isometric, or alternatively the cubic, system. More complex symmetry elements, such as inversion symmetry, glide symmetry and screw symmetry, are not often encountered in gemmology texts.

Crystallographic axes

Minerals are grouped into seven crystal systems having similar optical properties. Describing a crystal system and the shapes of crystals belonging to it is done with reference to crystallographic axes. Crystallographic axes are fixed lines of reference used to measure and describe the relative proportions of crystal faces and the angles between them. These axes are imaginary lines of specific length ratios running through and confined within a perfectly formed crystal in certain definite directions, in relation to the symmetry of the crystal, and which meet within the crystal at a central point termed the origin. Five crystal systems require three crystallographic axes and the other two crystal systems require four crystallographic axes. These axes must not be confused with symmetry axes. Crystallographic axes are concerned with length and angles, whereas symmetry axes are concerned with duplication of shape on rotation.

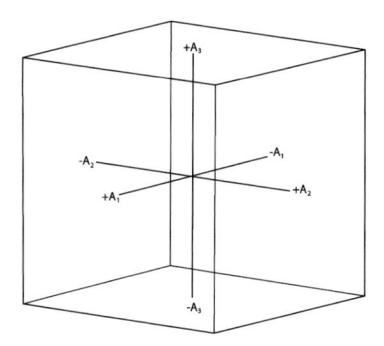

Figure 3.3 The crystallographic axes of the cubic system: three equal-length axes at right angles to each other produce equidimensional forms such as the cube. It is convention to number the front to back axis A1, the right to left axis A2, and the top to bottom axis A3.

The seven crystal systems

The relative lengths of the crystallographic axes and the angles at which they intersect define the seven crystal systems. The descriptions are of maximum symmetry; bear in mind that crystals from the same system but a lower class have lesser symmetry. The seven systems are presented in order of decreasing symmetry. All crystalline gems, regardless of crystal system, have directional physical properties – variations controlled by crystal structure – such as hardness, cleavage and grain. Physical properties relating to light are termed optical properties, such as refractive index, birefringence and pleochroism. Here is an important distinction: all crystals have directional physical properties, but only *anisotropic* materials – that is, those that are not amorphous and are from other than the isometric system – have directional optical properties. Light behaves the same in all directions in isotropic gems. This distinction is of primary importance in gem testing as many gems are differentiated by refractive index and birefringence. It is best to learn the systems in the order presented as this makes clear these optical properties and terminology for the future: the isometric system is *isotropic*; systems with equal horizontal crystal axes are birefringent and *uniaxial* – the tetragonal, hexagonal and trigonal systems; and systems with unequal or sloped lateral crystal axes are birefringent and *biaxial* – the orthorhombic, monoclinic and triclinic systems. Uniaxial and biaxial refer to optic axes. *Optic axes* are often referred to as directions of single refraction in a doubly refracting mineral, but this is incorrect in the strict sense of true optic science. It is more correct to say that an optic axis is a direction through a birefringent stone where light travelling in this direction shows zero birefringence. Uniaxial denotes one such axis; biaxial denotes two such axes.

The diagrams accompanying the seven crystal systems are illustrative of a few of the minerals and crystal forms from each of the systems. There is some question about the accuracy of form names for lower symmetry crystal systems. The naming of forms is now seldom used in the monoclinic and triclinic systems because the Miller indices of modern crystallography, an advanced concept beyond the scope of this discussion, provide a precise description of these forms without complicated terminology. In the triclinic system, every form will be either a single face or pairs of faces across a centre of symmetry. There is a discussion on crystal forms following the presentation of the seven crystal systems.

Table 3.1 Characteristics of the seven crystal systems

Crystal system	Crystallographic axes and designation	Angles	Maximum symmetry		
			Planes	Centre	Axes
Isometric/Cubic	$a_1 = a_2 = a_3$	All 90°	9	1	13 (3x4) (4x3) (6x2)
Tetragonal	$a_1 = a_2 \neq c$	All 90°	5	1	5 (1x4) (4x2)
Hexagonal	$a_1 = a_2 = a_3 \neq c$	3 at 120° and 1 at 90° to their plane	7	1	7 (1x6) (6x2)
Trigonal	$a_1 = a_2 = a_3 \neq c$	3 at 120° and 1 at 90° to their plane	3	1	4 (1x3) (3x2)
Orthorhombic	$a \neq b \neq c$	All 90°	3	1	3 (3x2)
Monoclinic	$a \neq b \neq c$	2 at oblique angle and 1 at 90° to their plane	1	1	1 (1x2)
Triclinic	$a \neq b \neq c$	All inclined	0	1	0

I Isometric system

Also called the cubic system, this has three crystallographic axes: all at right angles to one another and all of equal length. The maximum symmetry in this system is 9 planes of symmetry and 13 axes of symmetry: three 4-fold, four 3-fold and six 2-fold. This designation produces equidimensional forms such as the 6-sided cube, 8-sided octahedron and 12-sided rhombic dodecahedron. Isotropic.

II Tetragonal system

This has three crystallographic axes: all at right angles to one another, two of equal length, and one that can be longer or shorter. The usually unequal-length axis is set vertical and is known as the principal crystal axis; the equal axes are the lateral axes. The maximum symmetry is five planes of symmetry and five axes of symmetry: one 4-fold and four 2-fold. Common forms include a 4-sided prism; a bipyramid of two 4-faced pyramids that would be joined base to base if there were no intervening forms; and the pinacoid, a pair of diametrically opposite faces. Uniaxial: the principal crystal axis is the optic axis.

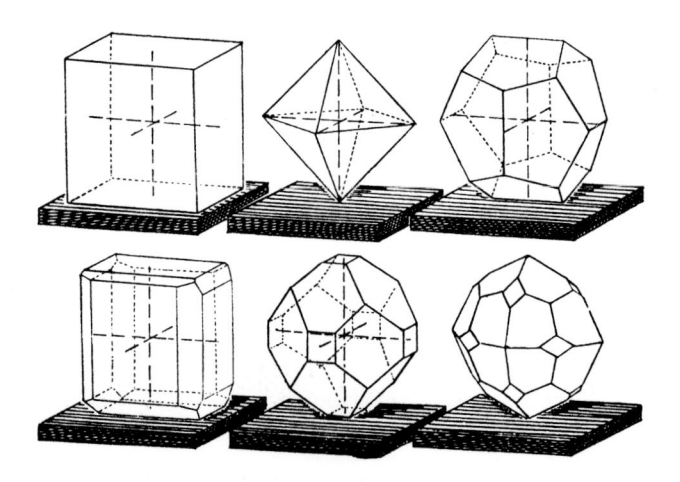

Figure 3.4 Crystal forms of the isometric or cubic system. Top row (left to right): cube – fluorite; octahedron – diamond; and pyritohedron – pyrite. Bottom row (left to right) combination forms: cube and pyritohedron – pyrite; cube and rhombic dodecahedron – garnet; and icositetrahedron and rhombic dodecahedron – garnet.

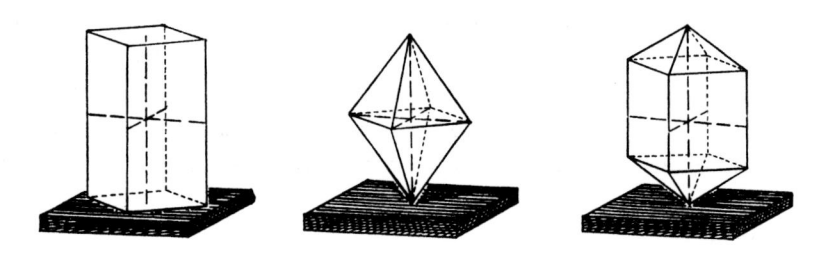

Figure 3.5 Crystal forms of the tetragonal system. Left to right: prism and basal pinacoid; tetragonal bipyramid; and combination of prism and bipyramid – zircon.

III Hexagonal system

This has four crystallographic axes: a vertical or principal axis that is usually longer or shorter than the other three, and the lateral axes, which are of equal length and intersect one another at 120°. The principal axis is at right angles to the plane of the lateral axes. The maximum symmetry is seven planes of symmetry and seven axes of symmetry: one 6-fold and six 2-fold. Common forms include a 6-sided prism and a bipyramid of two 6-faced pyramids that would be joined base to base if there were no intervening forms. Uniaxial: the principal crystal axis is the optic axis.

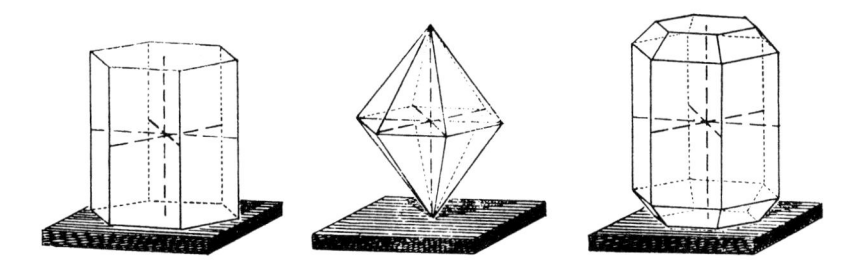

Figure 3.6 Crystal forms of the hexagonal system. Left to right: prism and basal pinacoid – emerald; hexagonal bipyramid; and combination of prism, bipyramid and basal pinacoid – beryl.

IV Trigonal system

This has four crystallographic axes: the same as in the hexagonal system, but the vertical symmetry is 3-fold rather than 6-fold. The maximum symmetry is three planes of symmetry, none of which are horizontal, and four axes of symmetry: one 3-fold and three 2-fold. The classic form is the rhombohedron comprising of six rhombus shapes, three around each end across a centre of symmetry. Uniaxial: the principal crystal axis is the optic axis.

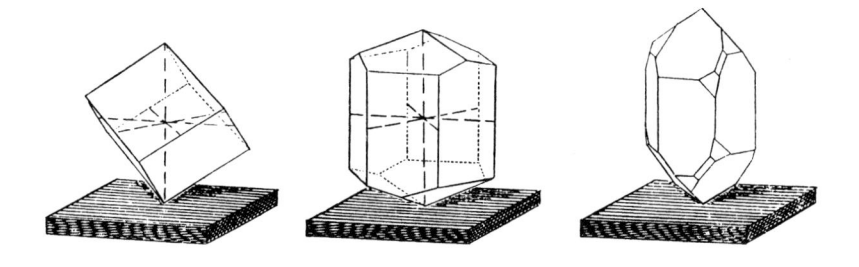

Figure 3.7 Crystal forms of the trigonal system. Left to right: rhombohedron – calcite; prism and rhombohedron – calcite; and hexagonal prism, major rhombohedron, secondary rhombohedron at alternating ends of the prism faces and two different trigonal trapezohedrons – quartz.

V Orthorhombic system

This has three crystallographic axes: all of usually unequal length and all of which are at right angles to one another. The main axis is vertical and the lateral axes pass through the sides; they are termed the *macro* axis for the longer of the two and the *brachy* axis for the shorter of the two. The maximum symmetry is three planes of symmetry and three axes of 2-fold symmetry. Common forms are a 4-sided prism with a rhombus-shaped cross-section like the 'diamond' on a playing card, and a bipyramid of 4-faced pyramids with oblong bases that would be joined base to base if there were no intervening forms. Biaxial: the two optic axes do not coincide with any crystallographic axes.

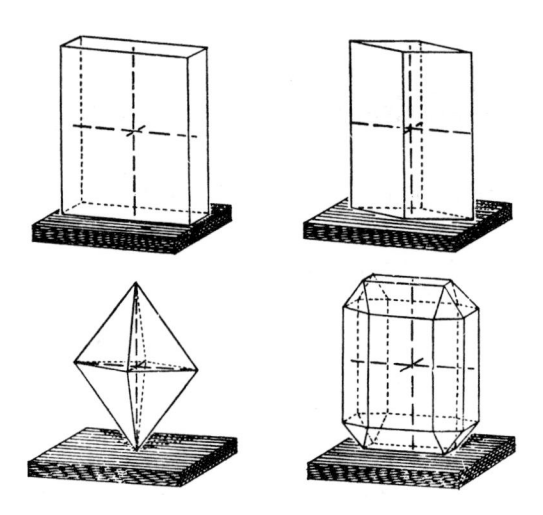

Figure 3.8 Crystal forms of the orthorhombic system. Top left: macro, brachy and basal pinacoids. Top right: prism and basal pinacoid. Bottom left: bipyramid. Bottom right: combination of three pinacoids – ortho, macro and brachy, three prisms – vertical, macro and brachy, and a bipyramid – peridot.

VI Monoclinic system

This has three crystallographic axes: all of usually unequal length, two of which intersect each other at an oblique angle, while the third is perpendicular to them. One axis is placed vertically and of the other two

CRYSTALLOGRAPHY

lateral axes, the one at right angles to the vertical axis is called the *ortho* axis and the axis that is inclined is called the *clino* axis. The maximum symmetry is one plane of symmetry and one 2-fold axis of symmetry. The shape may be best visualized as a prism on an oblong base, which has been pushed on one side, making it lean over slightly in one direction only. Forms include prisms, domes and pinacoids. Biaxial: the two optic axes do not coincide with any crystallographic axes.

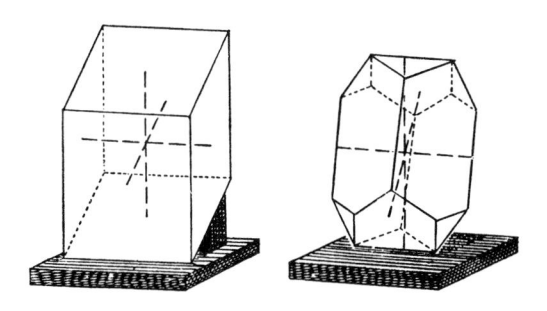

Figure 3.9 Crystal forms of the monoclinic system. Left: ortho, clino and basal pinacoids. Right: combination of prism, orthopinacoid, hemi-pyramid and the basal pinacoid – sphene.

Table 3.2 Minerals by crystal system

Isometric or cubic system	Tetragonal system	Hexagonal system	Trigonal system	Ortho-rhombic system	Monoclinic system	Triclinic system
Diamond	Zircon	Beryl	Corundum	Topaz	Jadeite	Kyanite
Garnet	Rutile	Apatite	Quartz	Peridot	Nephrite	Rhodonite
Spinel	Scapolite	Benitoite	Tourmaline	Andalusite	Spodumene	Turquoise
Fluorite	Idocrase	Sugilite	Calcite	Sinhalite	Orthoclase	Microcline
Pyrite	Cassiterite		Hematite	Chrysoberyl	feldspar	feldspar
Chromite	Scheelite		Dioptase	Iolite	Azurite	Plagioclase
Sphalerite			Phenakite	Danburite	Sphene	feldspar
Sodalite			Rhodochrosite	Enstatite	Malachite	Axinite
				Zoisite	Brazilianite	Amblygonite
			Smithsonite	Staurolite		

VII Triclinic system

This has three crystallographic axes: all of usually unequal length and all inclined to one another. One axis is placed vertically and the two lateral axes are called the macro and brachy as in the orthorhombic system. The shape may be best visualized as a prism on an oblong base, similar to that of the monoclinic system, but with the push given to an edge so that the prism leans over both backwards and sideways. Forms include pinacoids, hemi-prisms and hemi-domes with two faces each. There are no symmetry planes or axes. Biaxial: the two optic axes do not coincide with any crystallographic axes.

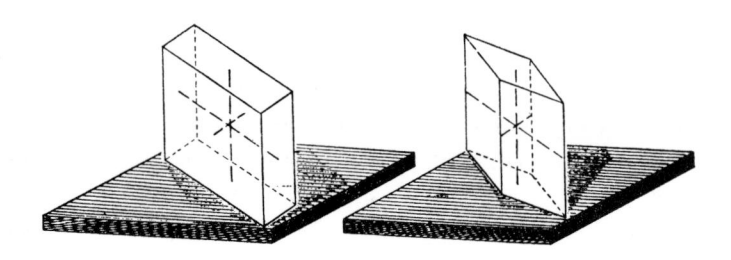

Figure 3.10 Crystal forms of the triclinic system. Left: macro, brachy and basal pinacoids. Right: basal pinacoid and hemi-prism.

Crystal forms and habit

A crystal form consists collectively of all those faces that are identically related to a crystallographic axis. There are many different forms. *Prisms* have three or more similar faces, all of which are parallel to the same axis – most usually the principal crystal axis. *Pyramid* forms consist of three or more non-parallel, typically triangular faces, capable of intersecting at a common point. A *bipyramid* has two pyramids base to base across a plane of symmetry. A *pinacoid* is a pair of parallel faces related by a centre of symmetry or a plane of symmetry. The *basal pinacoid* is the pair of faces parallel to the plane containing the lateral axes. A *dome* comprises two faces related by a plane of symmetry that intersects the vertical axis and one lateral axis. A *pedion* is a single face unrelated to any other. It should be understood that a pyramid may be described as one half of a bipyramid in the same way that a pedion looks like half of a pinacoid. Even though these are all different forms it is not always possible to differentiate between them. Consider a crystal in which it is not possible to see both terminations; in such a case one cannot distinguish a pyramid from a bipyramid or a pedion from a pinacoid. What is crucial to understand is that the importance is on overall symmetry: the bipyramid and pinacoid are of higher symmetry than the pyramid and pedion.

If the faces of a form can enclose space, or would do if extended, to exclude any other form, it is called a *closed* form. The cube is a closed form of six square faces. Every form of the isometric system is a closed form because the symmetry requires it. If a form cannot enclose space, even if extended, unless combined with another form it is called an *open* form. All prisms are open forms: the four vertical sides of a tetragonal prism require the top and bottom to be completed by the basal pinacoid to enclose space. The number of faces of a form depends on both the symmetry of the crystal class to which it belongs and the orientation of any given face to the symmetry axes or planes. Forms may be composed of 2, 4, 6, 8, 12, 24 or 48 similar faces. If the crystal is entirely made up of like faces it is termed a *simple* form. In all crystal systems, combinations of two or more forms are common; however, only forms belonging to the same class can occur together.

Crystals that have the full complement of symmetry elements possible for the particular system are referred to as *holohedral*. Within the same system if a crystal lacks a centre of symmetry it belongs to a class of lower symmetry, one with half of the symmetry of the holohedral class, and is referred to as *hemihedral*. All well-formed crystals are referred to as *euhedral*. Crystals with different forms at opposite ends of an axis of symmetry, generally the vertical axis and therefore these are different terminations, are known as *hemimorphic*. Such crystals often show electrical effects, as does tourmaline. Through inequalities of growth, crystals rarely conform to the ideal shapes presented here but no matter how distorted a crystal is, it is always possible to find some faces that help identify the forms present. A good rule of thumb is that faces with the same growth features will belong to the same form. It is crucial to understand that the term 'form' is not a synonym for 'shape'. Form has a precise technical meaning. The general term for the shape, or outward appearance, of a crystal is *habit*.

A great deal of confusion can arise with the term habit because the term is used in two distinct senses. It can refer to an individual crystal at hand, in which case habit is a description of the individual crystal shape including its form(s), proportions and growth marks. In this instance the term differentiates between individual specimens of the same species, and habit may vary from one crystal to another. Also, habit can refer to a mineral in general, in which case habit is the characteristic shape of a particular mineral relating to one or more forms and their relative proportions. This is habit as a descriptive property of a mineral on the whole, and is an established physical property. Habit in this instance is often described in general appearance terms such as prismatic, which means with vertical faces well developed, as describes tourmaline. Diamond is an exception; it has octahedral habit as it is found most often as octahedra.

Slender needle-like crystals, such as rutile in quartz, have *acicular* habit. The branch-like inclusions such as those seen in moss agate are known as *dendritic*. Crystals forming spherical or hemi-spherical masses produce the so-called *globular* or *botryoidal* habit that is common in malachite. *Granular* material consists of an aggregate of large or small grains and is exemplified by quartzite and marble. *Stalagmitic* material is formed by deposition of minerals from water solution and may form stalactites, which are pendant cylinders from the roof of caves, or stalagmites, which are the converse of stalactites in that they grow upwards from the floor of the cave below the stalactite. If they join they form a column. A mineral whose external habit is not the one usually associated with its particular species is a *pseudomorph*. The original material, which may have been organic or inorganic, has been replaced particle by particle with another mineral but it usually keeps its original habit.

Habit often varies in crystals of the same species but from different localities, or by different varieties of the same species. In the corundum species, the habit of ruby is tabular and the habit of sapphire is bipyramidal; however, some sapphires such as those from Montana are of short prismatic or even tabular habit.

Twinned crystals

Twinned crystals consist of two or more components that have grown together in a symmetrical manner and that have a direct crystallographic relationship to each other, most commonly with one part in reverse orientation to the other. It is often reported that twin crystals 'consist of two or more individuals...' but this is incorrect. Twins are not two or more individual crystals. They are single individual crystals that have grown as individual crystals from the start; however, they have growth disruptions between two or more components.

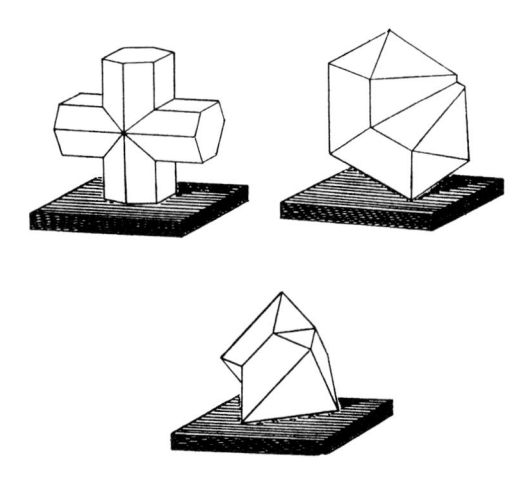

Figure 3.11 Twinned crystals. Top left: a staurolite twin, the so-called 'cross stone' from the orthorhombic system. Top right: a geniculate, or knee-shaped, twin characteristic of zircon from the tetragonal system. Bottom: the macle, a contact rotation twin, or spinel twin, characteristic of diamond from the isometric system.

A *contact rotation* twin is one in which the components are in contact along a common plane such that if one half were rotated 180° about an axis, the twinning axis, it would bring them both into parallel position. That is to say, it would reform the two halves so as to make a single untwinned crystal. This type of twinning is typical of spinel, and is often called spinel twinning. It produces the *macle*, the triangular contact rotation twin characteristic of diamond. *Interpenetrant* twins are those in which the two components have intergrown such that they appear to penetrate one another. This type of twinning produces cross-like twins such as those of staurolite. *Polysynthetic* twinning, also known as *repeated* or *lamellar* twinning, refers to repeated twinning of components in a plane. The components are often very thin plates stacked together like cards in a deck except that the atoms of alternate plates are in reverse orientation such that the deck has every second card rotated 180° with respect to the one above and below it. The components may be so thin they appear as striations. This type of twinning can occur after crystal formation as a result of extensive heat and pressure that distorts the structure through the flat layers of the crystal. A twinned crystal typically shows a re-entrant angle, likened to an inside corner. This can distinguish it from parallel growth in which groups of crystals join together with the faces and edges parallel to those of the other crystals; in this case the faces on one side are parallel to the faces on the other side of the angle.

LESSON 4

Hardness, Cleavage, Parting and Fracture

Durability, one of the three cardinal virtues of a gemstone, is the ability to withstand wear, damage or decay. In gemmology, hardness, cleavage, parting and fracture are all influences to one degree or another on the durability of a gemstone. When a mineral breaks it does so by fracture (ranging from large cavities and chips to small nicks), cleavage or parting. Toughness is the ability to withstand disruptive stress, particularly to resist fracture, parting and cleavage. Stability is also often mentioned, as it relates to a gemstone's ability to resist damage or decay due to heat, light or chemical attack. In general, organic gemstones are far less resistant to these elements than mineral gemstones. For mineral gemstones, the properties of hardness, cleavage, parting and fracture depend on the phenomenon known as cohesion. Cohesion is the adhesive attraction between molecules. Hardness depends on the strength of the bonds; and fracture, parting and cleavage all depend on the positioning of atoms and how the mineral breaks when force is applied.

Hardness

The term 'hardness' has many interpretations and there is more than one way to express the hardness of a material. In everyday language we characterize certain materials as hard, such as concrete, or soft, such as Styrofoam. In materials science, hardness refers to various properties of solid matter that give it resistance to different types of shape-change in response to different types of force. It is tested by various means as the resistance to: (a) deformation; and (b) abrasion, which includes scratching. Gemmologists are concerned almost exclusively with scratch hardness, and in gemmology hardness may be defined as the power a substance possesses to resist being scratched when a pointed fragment of another substance is drawn across it without sufficient pressure to cause breakage, fracture or cleavage.

Hardness is not attributable to the strength of individual atoms. We may note that carbon, which forms the incredibly hard diamond, also forms the extremely soft mineral graphite. Hardness is dependent upon the manner in which atoms combine: strong intermolecular bonds characterize harder materials. Diamond's atomic structure consists of carbon covalently bonded, the strongest form of attraction between chemical elements.

Mohs scale of hardness

The Mohs scale of hardness, devised by German mineralogist Friedrich Mohs in 1812, characterizes the scratch resistance of various minerals. Mohs proposed a series of ten readily available minerals with reasonably constant hardness and set them in order, softer to harder. This is not a

Table 4.1 Mohs hardness scale

The ten common minerals as ordinal standards	
Diamond	10
Corundum	9
Topaz	8
Quartz	7
Orthoclase feldspar	6
Apatite	5
Fluorspar	4
Calcite	3
Gypsum	2
Talc	1

linear scale comparable with the 10 millimetres in a centimetre; the intervals between the numbers are not equal. In fact, the difference in hardness between diamond (10) and corundum (9) is greater than that between corundum (9) and the softest mineral, talc (1). It is simply an ordinal scale that sets topaz as softer than corundum and fluorspar as harder than calcite. A fingernail (2.5) will scratch the two softest minerals on the scale, talc and gypsum, and those softer than quartz may be scratched by a steel knifepoint (6.5). There is no quantitative significance to the numbers; the test is merely based on the fact that any mineral of a certain established Mohs number will scratch minerals of equal or lower hardness numbers and be scratched by minerals of equal or higher numbers.

Scratch hardness testing

Scratch hardness testing may be conducted with hardness pencils or hardness plates. With hardness pencils (or hardness points, as they are sometimes known), the test is performed by observing if the sharp edge or point of a mineral of known hardness scratches a smooth flat surface of the mineral under test. For this purpose, pointed fragments of minerals numbered 2 to 10 are mounted in wooden or metal stems. It is possible to use synthetic spinel (8) and synthetic corundum (9), as synthetics are understood to have the same hardness as natural minerals. Working up from the softest to the hardest, the hardness pencils are applied to the test mineral. It is not necessary to make a long mark, just begin to move the point across the surface and feel if it is biting into the test mineral or whether it is simply sliding over the surface with no penetration. As small a mark as possible should be made. All marks should be wiped to observe whether or not the test mineral was in fact scratched or whether the point of established hardness was abraded by the procedure, leaving a removable dust trail. Hardness pencil tests may be useful for rough crystals and sometimes for statuary and carvings where a tiny mark made on the base or underside will not detract from its value – but this is a destructive test and should be avoided with cut gemstones. If it absolutely must be performed, test the girdle as it is often unpolished, or a pavilion facet edge. The test stone's hardness will be between that of the point that does not scratch up to, or equal to, that of the point which does.

A less destructive method is to use hardness plates; these are small polished plates of minerals of known hardness (or their synthetic counterparts) against which an edge of the test stone can be drawn. This way, the mark is made on the plate and not on the test stone. The hardness of the test stone will be less than the plate it does not scratch, and greater than or equal to the plate it does scratch. Be certain only to use moderate pressure, as it is possible to develop cleavage or cause the stone to fracture or chip with excessive force.

This is important to understand: a test mineral's hardness may be equal to that of the plate it scratches or, in the case of hardness pencil tests, equal to the hardness point that scratches. It is often incorrectly reported that nothing scratches corundum or a plate of synthetic corundum except diamond, but corundum scratches corundum and so does the newest diamond simulant, synthetic moissanite. One of the oldest tests for diamond was whether or not it would scratch glass. This test is often used for dramatic purposes in films but of course there are many other gems and imitations capable of scratching glass, although it should be noted that they do not do so nearly as deeply or as easily as diamond.

Scratch hardness tests are unnecessary. Gemmologists have an obligation to return the specimens they are entrusted with undamaged, even imitations. A test that was often resorted to in the early twentieth century, it is now rarely used, as the science of gemmology has evolved considerably and more modern and safer scientific methods have adequately replaced the hardness test.

Although it is unacceptable to test for it directly, hardness can still be of great value in gem identification because it can be assessed visually. With the loupe or microscope it is possible to see many of the characteristics indicating gemstone hardness, including: wear on facet edges, scratches, chips, the flatness and polish of facets, and the sharpness of facet edges. This is often what gives away glass imitations on initial inspection: they are soft, and typically show rounded or moulded facet edges and considerable signs of wear. Dust contains microscopic particles of sand, which are pulverized quartz (7). Rubbing a gemstone on one's sleeve or wiping it with a cloth to clean it grinds these particles across the surface in an abrasive action that will eventually destroy beauty and lustre. Glass imitations have hardness of less than 6 and are easily damaged in this manner. With glass, in order to increase its lustre and brilliance to make it a more effective simulant, small quantities of the oxides of lead and thallium must be added, both of which soften it considerably.

Factors affecting hardness

A number of factors affect hardness, principally: twinning, impurities and direction. It is found that stones from different localities may vary slightly in hardness. Diamonds discovered in New South Wales, Australia and those from Borneo are found to be harder, or at least harder to cut, than stones from South America, which are in turn harder than those from South Africa. There is no fundamental difference in the physical hardness of diamonds from one location or another; the apparent difference is attributable to knots. Knots slow the polishing process such that when cutting and polishing time is taken as a measure of hardness, knotted stones seem harder. The grain is the key. The plane on which crystals twin, or where interpenetrant twins have grown together, presents a reversal of grain, termed cross-grain, an area highly resistant to polishing.

Impurities also affect hardness, and since impurities are responsible for colour in many gemstones, colour differences often relate to hardness differences. In the corundum species all sapphires are understood to be somewhat harder than ruby. With diamond, nitrogen content on the order of a few nitrogen atoms per 100 000 carbon atoms produces a yellow colour and also strengthens the crystal.

Hardness may also vary with direction; in particular this is noticeable with kyanite as it shows hardness of 5 along the length of the crystal and 7 across the crystal. Diamond also shows a variation in hardness with direction that can exceed 100:1. Directional hardness variations are directly attributable to the symmetry of the crystal. A general principle of minerals is that they are typically harder in a direction across the edge of a cleavage plane. The fact that the Mohs scale does not account for directional hardness is one of its major limitations; the other is that being ordinal it is non-linear and non-quantitative.

Theoretically, naturals and synthetics are equal in hardness, but in reality there may be some exceptions. Both natural and synthetic corundum are 9, but the synthetics are usually found to be slightly harder. However, hardness cannot be used to differentiate between naturals and synthetics because these differences are insignificant for testing.

Gemmological importance

The property of hardness is important in gemmology for a number of reasons. Generally, the harder a mineral is the higher the polish it will take. Sapphire (9) takes a better polish and exhibits a higher lustre than topaz (8). Diamond is the hardest mineral and has the best lustre of all, referred to as adamantine. Also, harder minerals exhibit sharper edges and flatter facet surfaces when faceted. In fact, this is one of the ways an experienced gemmologist comes to identify diamond. In terms of faceting, the harder a mineral is, the longer it will take to cut, and therefore the more expensive it will be to fashion. Directional hardness dictates the allowable cutting orientations. Diamond owes the fact it can be cut at all to the existence of directional hardness. Diamond is only cut by its own dust in which some of the randomly oriented grains have their harder directions applied to a softer direction in the cutting stone. Stone setters are interested in hardness because they need to know how much pressure the stone can take without damage. Consumers need to understand proper care and how well the stone will stand up to everyday wear and so retain its beauty and value. Never toss gemstones together in a reckless manner. All fine stones should be stored separately to prevent them from coming into contact with one another. This is particularly important where diamond is involved, as it readily does harm to other softer gemstones.

Testing diamond hardness

Hardness tests based on the non-linear Mohs scale cannot properly express the extreme hardness of diamond. This requires more precise evaluations with more exact measurements. Scratch hardness can be measured with greater precision by an instrument known as a sclerometer. Testing involves mounting the test material with its surface horizontal and drawing a very sharp diamond point across this with increasing pressure until the diamond scratches the surface. When tested in this way, the hardness attributed to quartz is 245, corundum 1000, and diamond 140 000.

Abrasion hardness tests typically employ techniques subjecting the test material to a grinding wheel charged with diamond powder for a specified time interval and then measuring the amount of material removed. E. M. Wilks at Oxford University tested the relative hardness of diamond and sapphire and found that when tested in the hardest direction, sapphire wears 5000 times more readily than diamond.

Table 4.2 Hardness

Hardness	Mineral
10	Diamond
9	Corundum
8.5	Chrysoberyl
8	Spinel
	Topaz
7.5	Almandine garnet
	Andalusite
7.25	Hessonite garnet
	Pyrope garnet
	Rhodolite garnet
	Spessartite garnet
7 to 7.5	Iolite
	Tourmaline
	Zircon
7	Jadeite
	Quartz
	Spodumene
6.5	Benitoite
	Demantoid garnet
	Nephrite
	Peridot
	Sinhalite
6	Orthoclase feldspar
	Scapolite
5.5	Sphene
	Lapis Lazuli
	Enstatite
5	Apatite
	Obsidian
4	Rhodochrosite
2.5	Ivory
2 to 2.5	Amber

Cleavage

Cleavage refers to the tendency of a crystalline substance to split parallel to certain definite directions, in response to force, leaving a basically smooth flat surface. The property is likened to the tendency of wood to be split along the grain. Cleavage is related to symmetry and is due to the presence of planes of weakness created when the bonding strength between layers of atoms is not the same in every direction. In a crystal structure, atoms are arranged in regular repeating patterns; therefore layers of atoms exist parallel to certain definite crystal directions. Some of these layers will have greater bond strength than others and cleavage refers to the tendency of a crystalline mineral to split along the weaker planes. The preferred planar direction in which a crystal will part under force or stress is referred to as the cleavage direction. Cleavage is a directional property related to crystal symmetry, and therefore cleavage can occur in any plane parallel to the cleavage direction. Cleavage only occurs with crystalline substances. Amorphous substances fracture in random directions. Cleavage is always parallel to a possible crystal face, but this face is not always formed. Directions of cleavage remain the same regardless of the external shape of the crystal. Even when the gemstone is a water-worn pebble, highly distorted rough or has been faceted, cleavage still occurs in the same directions. Cleavage is a characteristic property: the same mineral always has the same cleavage.

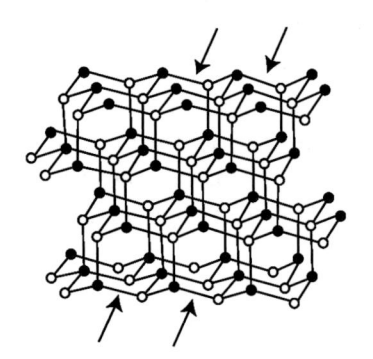

Figure 4.1 Cleavage planes in diamond.

With diamond, even though the strong atomic bonds are the same length in every direction, when this forms a three-dimensional structure, it is configured such that fewer bonds cross some planes. These planes, which are said to have lower bond density, are the planes on which the crystal is weaker and therefore susceptible to cleaving. Diamond has octahedral cleavage because it is the octahedral plane that is layered with the greatest and least massing of bonds; therefore, alternate atomic layers in this direction are held together by relatively few bonds, allowing diamond to cleave at the layer between the two closely bonded layers. Each direction of cleavage means that the cleavage will exist on the two opposite sides of a three-dimensional structure. Diamond has four cleavage directions because the eight faces of the octahedron are four parallel pairs.

There is no single scientifically recognized categorization for cleavage. Typically it is classified in three ways: (1) the quality of the cleavage surface; (2) the direction of cleavage; and (3) the ease of developing cleavage. In terms of the quality of the cleavage surface it is described as: *perfect* – a cleavage surface without any rough surfaces, leaving an extremely smooth flat plane such as with diamond and topaz; *good* – still a smooth surface but with slight roughness, ripples or markings such as with feldspar; *fair* – still noticeably flat but smooth only in small areas such as with scheelite; *poor* – leaving a flat rough surface dominant such as with apatite; and *none* – meaning that the mineral exhibits no cleavage such that if it breaks it leaves an irregular surface as a fracture. Gemstones noted for their absence of cleavage include: corundum, garnet, spinel, quartz and tourmaline. *Indistinct* is often used to refer to a mineral that exhibits cleavage, but the cleavage is poor and hardly noticeable – such as with beryl.

Cleavage is sometimes described in mineralogy by its number of directions: one, two, three, four and six directions. In gemmology, it is typical to name the direction. Topaz and beryl have basal cleavage; pyrite has cubic cleavage; spodumene has prismatic cleavage; and olivine has pinacoidal cleavage.

Cleavage is also described in terms of the ease with which it can be developed. In this manner, cleavage can be described as: easy, moderate or difficult. Fluorite and diamond both have perfect octahedral cleavage, but fluorite cleaves easily and diamond with difficulty. The use of these terms is often inconsistent and readers may find differences from source to source. The importance is understanding the meaning, not the particular term itself.

Since cleavage, or absence of cleavage, is characteristic of a mineral, it is sometimes helpful in identification. Cleavage is a particularly important characteristic with crystals as it can help to identify the mineral's symmetry. Since cleavage must obey the symmetry of the crystal, crystallography dictates that a mineral belonging to the isometric system can exhibit either three or four directions of identical cleavage or no cleavage at all. Uniaxial minerals, those from the tetragonal, hexagonal and trigonal systems, typically have basal cleavage perpendicular to the major crystallographic axis or prismatic cleavage of respectively four, six, or three directions parallel to the length of the crystal. Biaxial minerals, those from the orthorhombic, monoclinic and triclinic systems, can never exhibit more than two identical cleavage directions.

Cleavage can be seen without the mineral being cleaved all the way through. It is often shown as an internal reflection plane exhibiting rainbow interference colours. Topaz is easily cleaved in the basal direction and it is often possible to see internal evidence of cleavage in a faceted topaz.

Cleavage is advantageous in gem cutting as it is often used on large diamonds to bring them into a more regular shape without time-consuming sawing. All cutters must understand cleavage, as it is not possible to polish a gemstone on a cleavage plane and achieve a smooth surface; so this must be taken into account when planning to cut a cleavable gemstone. Cleavage can also be disadvantageous in that if an easily cleavable stone is dropped there is a greater likelihood for it to break in two or, at least, if this greater misfortune does not occur, for the drop to induce internal flaws that can destroy transparency and create zones of weakness that may be more easily extended.

Parting

Parting is characteristically similar to cleavage, and for this reason is often referred to as false cleavage or pseudo-cleavage. In spite of these terms, it must not be confused with cleavage. Parting is a separate phenomenon. It occurs in minerals that do not exhibit cleavage but can be divided into two along a plane of weakness, leaving a flattish surface. The direction of weakness is called the parting plane and twinning usually causes it. It is characteristic of corundum, which parts along the directions of lamellar twinning in which the twins are very thin plates stacked together like a deck of cards. It is just a zone of weakness between twinned layers; it may be parallel to a possible crystal face but does not occur in all planes parallel to this direction – it only occurs at the planes joining twin components. Parting is far less common than cleavage.

Fracture

Fracture refers to a random, non-directional breakage, occurring in any direction other than a cleavage, usually as a result of sharp impact. All minerals fracture, even those that do not exhibit cleavage. Many minerals exhibit both fracture and cleavage. Toughness and brittleness are related concepts. Toughness is the ability to withstand breaking either by fracture, parting, cleavage or a combination. Brittleness is the readiness of the gemstone to fracture. Sometimes a relatively hard gemstone, such as zircon (7 to 7.5), is brittle and, in spite of its relative hardness, it readily chips. With zircon the brittleness is a result of the heat treatment process to enhance colour. On the other hand, some minerals with moderate hardness, such as jadeite (7) and nephrite (6.5), are very tough and hardly ever fracture because their structure is interlocking fibres or crystals, a structure characteristically tougher than that of a single crystal.

Fracture is characterized by the appearance of the fractured surface. Almost any verbal description can describe a fracture, especially:

- *Conchoidal* – meaning shell-like, because it breaks leaving curved concavities resembling a shell. This type of fracture is typical of quartz, garnet and glass. Diamond fractures conchoidally and it often occurs in combination with cleavage, leaving a surface with a step-like pattern.
- *Smooth or even* – when the surface, without being absolutely plane, presents no marked irregularities. This is sometimes seen in opal.
- *Splintery* – when the surface is covered with partially separated splinters in irregular fibres. This fracture is common with chalcedony, jadeite and nephrite – all fibrous minerals exhibit this type of fracture.
- *Hackly* – when the surface is covered with rough, jagged points and depressions. All true metals exhibit this type of fracture.
- *Uneven* – leaving a rough or irregular surface such as with corundum.

Other categories often include *crumbly*, when the mineral crumbles when fractured; *granular*, when the surface is like a broken cube of sugar; and *subconchoidal*, referring to a break that is smooth with irregular rounded corners but not quite conchoidal.

Almost all minerals have a characteristic fracture. Occasionally it can be useful for identification. The vitreous lustre on a conchoidal fracture is often what helps to distinguish a glass imitation. Fractures are rarely found on minerals with easy cleavage because when force is applied the mineral tends to cleave rather than fracture. It is more common on minerals that exhibit no cleavage or cleave with difficulty.

LESSON 5

Density and Specific Gravity

Which weighs more, a pound of lead or a pound of feathers? It is one of the older riddles which, absurd as it apparently is, concerns a vital point: the relationship of weight to volume. For the space taken up by the pound of lead is far smaller than the space taken up by the pound of feathers, even if they were compressed to their smallest bulk. Likewise, the comparatively heavy metal iron may be said to weigh more than the light metal aluminium, but if 2kg of aluminium and 1kg of iron are compared, it is obvious that the aluminium would be twice as heavy as the iron. But, if equal quantities by size, such as 1cm cubes, were compared it would then be found that the iron cube would weigh approximately three times as much as the cube of aluminium. Thus, it is said that the *density* of iron is about three times that of aluminium.

The relative weight, or heaviness, of any material is one of its fundamental properties. In everyday life we think of certain materials as light, such as cork, and others as heavy, such as lead; however, for the comparison to have any scientific significance it is necessary to consider the volume of each object compared to the other. The formal definition of density is *mass per unit volume*. In everyday language the terms 'mass' and 'weight' are used interchangeably, but in science there is an important distinction. Mass is a measure of the quantity of matter and is constant, whereas weight is the force exerted on a body by gravitational attraction. Weight is proportional to mass, but it varies in different parts of the universe. A man weighing 90kg on Earth is weightless in space and would weigh 15kg on the moon where the force of gravity is one-sixth of what it is on Earth.

Density is calculated as the mass of a substance in grams divided by its volume in cubic centimetres (g/cm³), alternatively expressed as kilograms per cubic metre (kg/m³), grams per millilitre (g/ml) or kilograms per litre (kg/l). With the units g/cm³ and g/ml the numbers will be the same, as a cubic centimetre is equal to a millilitre. And with kg/m³ and kg/l the numbers will be greater because a litre and a kilogram are 1000 times greater than a millilitre and a gram. Density is expressed most commonly in g/cm³ or kg/m³. The density of aluminium is 2.7g/cm³ and the density of iron is almost three times that of aluminium at 7.8g/cm³.

The mean value for the density of diamond is 3.515g/cm³ or 3515kg/m³; however, gemmologists refer to another measure of density, that of *specific gravity*. Specific gravity (SG) is defined as the ratio of the density of a material compared to an equal volume of pure water at 4°C at standard atmospheric pressure. Specific gravity is a term used for solids. The term *relative density* is reserved for comparing one liquid with another although in gemmology this distinction is seldom made and the term 'specific gravity' is used in both cases. Water is the chosen standard as it is stable, inexpensive and readily available. The specific gravity of water at 4°C, its maximum density, is by definition fixed at 1.00. For all practical purposes it is not necessary to correct for the temperature of the water. Water at room temperature, 20–25°C, is near enough at 0.99820–0.99704g/ml to unity (1.00) to have no realistic effect on test results.

Specific gravity and relative density are ratios expressed without units. Although they are often termed 'density' for short, correctly speaking, density measurements require units. Density is a physical property of all materials. Every substance on Earth has a particular density and as a rule this density is constant between fairly narrow limits for pure substances. Gemstones are mostly pure substances and their specific gravities are, with few exceptions, remarkably constant, hence the property may be used as a means for their identification. The specific gravity of diamond is 3.515 rounded to 3.52 for practical purposes, meaning that diamond is 3.52 times heavier than an equal volume of water. Materials consisting of elements with low atomic weights such as diamonds generally have lighter densities, but diamond is an exception because the more closely packed the atoms are, the higher the SG. The carbon atoms in diamond are bonded very closely, making its tight compact structure relatively heavy for its atomic weight.

In reality, gemmologists seldom use specific gravity tests. In many cases it is a test of last resort because it is seldom sufficient to identify a gemstone. It cannot distinguish between natural and synthetic gemstones and has several other drawbacks depending on the chosen test method. The main gemmological tests are optical, not physical, so specific gravity tests are secondary tests; however, in instances where they are called for they provide accurate information to help identify unmounted gemstones.

At first it would appear to be an insurmountable difficulty to calculate the volume of a gemstone, which is geometrically complex in shape, in order to compare it with an equal volume of water. Fortunately, Archimedes solved this problem in the third century BC.

Archimedes' Principle

The Greek mathematician and physicist Archimedes is credited with discovering the principle and method to determine the specific gravity of an object. He was asked to determine whether the King's crown was solid gold or whether the goldsmith had been dishonest and mixed in some silver. If the crown could have been melted down to a regularly shaped body, such as a cube, it would have been an easy task to calculate its volume for comparison to the weight of an equal volume of water, but obviously the crown could not be damaged. Legend has it that the solution came to Archimedes while taking a bath. It happened that his bath was more filled with water than usual and, as he got in, two things occurred. As he lowered himself he experienced the floating sensation of water action upon his body, and at the same time his bath overflowed. In considering these two points and the relationship between them, he had the great insight that the upward 'flotation' force he experienced must be equal to the weight of the volume of water he displaced and, furthermore, if the tub had been filled to the brim before he got in, he could have caught all the overflow and easily measured the volume of the displaced water. He is said to have jumped from his bath shouting 'eureka' when he reached this understanding. He realized he could use this principle to determine the volume of the crown, although it is not known which of three possible approaches he used.

The first approach would be to suspend the crown on a string from a weigh scale and measure its weight in air, and then lower the scale and crown so that the crown (but not the scale) became immersed in water and he could measure the weight of the crown while it was still immersed in water. The simple calculation of weight in air divided by the loss of weight by immersion in water will give the specific gravity of the metal. Knowing that the specific gravity of gold is 19, if the calculation were less than this it would prove the crown was not pure gold and that the lighter metal silver was present.

The second approach would be a direct comparison without calculation. Here a simple balance scale would be used to compare the crown balanced against an equal mass of pure gold. The balanced scale would then be fully immersed in water. If the crown and the lump of gold were in balance under the water, then the crown must be fine gold because both experience identical upward force. If the crown came to rest higher than the gold it would indicate that the crown had a lower SG because it had displaced more water due to its larger volume, and therefore experienced a greater upward force, causing it to rest above the gold.

The third approach he could have used would have been the following: to fill a vessel with water, catch the overflow when the crown is immersed, and then weigh the displaced water and divide the weight of that water into the weight of the dry crown to calculate the SG of the crown. This type of overflow vessel is named a *eureka can* in honour of the moment that Archimedes' bath overflowed and prompted him to arrive at his understanding that has since been known as Archimedes' Principle. The methods of specific gravity measurement applied in gemmology follow from Archimedes: hydrostatic weighing, water displacement and heavy liquids.

Hydrostatic weighing

The hydrostatic weighing method, alternatively termed the *direct weighing method*, depends on Archimedes' Principle that a body immersed in a liquid is buoyed up by a force equal to the weight of the liquid it displaces. Therefore, if a gemstone is first weighed in air, and then again when immersed in water, it will weigh less in the water by an amount equal to the weight of the water it displaces.

This method requires the use of an accurate balance or scale capable of weighing to a milligram. The procedure traditionally described is that employing a two-pan balance; however, single-pan balances, spring balances and specialized balances that give direct specific gravity readouts are also used.

DENSITY AND SPECIFIC GRAVITY

Using a two-pan balance, obtain a stool that can be used to stride the left-hand pan in such a manner that it does not in any way hinder the swing of the balance. Place a beaker of distilled water on the stool, again ensuring that it clears the pan arms of the balance. From the hook at the end of the beam, suspended by fine-gauge monofilament nylon, attach a cage of fine wire to hold the stone. This cage may be formed into a cone shape by winding the wire around the sharpened end of a pencil. The cage must be completely submerged in the water and remain that way throughout the full swing of the balance. On the hook over the right-hand pan, place a piece of wire exactly counterpoising the cage on the left-hand side to level the balance when the cage is immersed in the water. The stone under test may now be weighed in air by placing it in the left-hand pan and adding weights to the right-hand pan. Older balances came with a boxed set of weights in various increments; modern units have a rider that is slid along the graduated arm to obtain readings. If using an older model, follow a systematic procedure with the weights starting in excess of the gemstone and work downwards. Once the weight in air is obtained (X), the stone is removed from the left-hand pan and placed in the wire cage suspended in the water. The stone is now weighed in a similar manner and the weight, which will be less by a certain amount, is noted. This is the weight of the stone when immersed in water (Y). The apparent loss of weight (X − Y) when divided into the weight of the stone in air gives the specific gravity of the stone by the formula:

$$\text{Specific Gravity (SG)} = \frac{\text{weight of gemstone in grams}}{\text{loss of weight in water}} \text{ or } \frac{X}{X-Y}$$

If a red gemstone weighs 6.00ct in air (X) and weighs 4.50ct when immersed in water (Y), this gives an apparent loss of weight (X − Y) of 1.50ct. So the SG is calculated as 6.00 ÷ 1.50 = 4.00. The gemstone with an SG of 4.00 is corundum, and the red variety is ruby.

Air bubbles adhering to the stone can buoy it up and so cause erroneous readings. Carefully remove all air bubbles with a small camelhair paintbrush. The use of boiled or distilled water reduces the presence of bubbles. Calculations are taken only to two decimal places because drag on the wire, due to the surface tension of the water, precludes accuracy beyond the second decimal place. Adding a drop of dishwashing detergent to the water will effectively reduce surface tension.

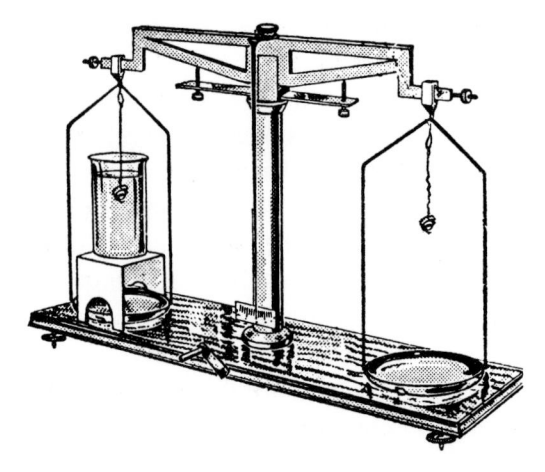

Figure 5.1 A balance set up for hydrostatic weighing.

Hydrostatic weighing is a non-destructive, simple procedure that is accurate for gemstones over 3ct. It is also useful for statuary and carvings. Its disadvantages are that it cannot be used with mounted stones, it is time-consuming, and the margin of error increases in inverse proportion to the size of the gemstone. Suitable scales can be relatively expensive.

Water displacement

It is unnecessary to ascertain a gemstone's volume by mathematical calculation to determine its specific gravity, for if the stone is placed in a eureka can it will displace a volume of water equal to its own volume. The eureka can is a metal vessel fitted with an overflow pipe, which is filled with water until it just starts to overflow. Once the water level settles with the can filled to the brim, the gemstone is gently placed in the can, causing the water level to rise, and the displaced water passes through the overflow pipe and can be collected and measured. Thus, if the displaced water is measured in cubic centimetres and the stone is weighed in grams, all that is necessary is a simple calculation because 1 cu cm of water at 4°C weighs exactly 1g: divide the weight of the gemstone by the volume or weight of the water. Either formula yields the answer:

$$\text{Specific Gravity (SG)} = \frac{\text{weight of gemstone in grams}}{\text{volume of displaced water in cubic centimetres}}$$

DENSITY AND SPECIFIC GRAVITY

$$\text{Specific Gravity (SG)} = \frac{\text{weight of gemstone in grams}}{\text{weight of displaced water in grams}}$$

If a purple gemstone weighs by scale 5.30g and displaces 2.00 cu cm of water, SG is calculated as 5.30 ÷ 2.00 = 2.65. The gemstone with an SG of 2.65 is quartz, and the purple variety is amethyst. Keep in mind that SG values are not always exact – if the stone has impurities or comes in a variety of colours, the value may vary slightly.

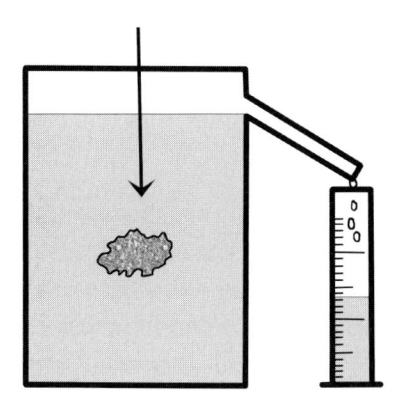

Figure 5.2 The eureka can. The object is placed into the can and a volume of water equal to the volume of the object is displaced through the overflow pipe.

Easy as this method is, it is only suited to larger crystals, carvings or *objets d'art*. In the previous example, one may come across a 26.5ct amethyst (5.30g), although most gemstones are far smaller than this, and this method does not lend itself to the accuracy required for small gemstones.

If it is necessary to determine the SG of a smaller gemstone by water displacement, the vessel used is a specific gravity bottle, the *pycnometer*: a small glass flask fitted with a ground glass stopper, which is pierced lengthwise by a capillary opening. It is generally engraved on the outside with its weight when empty, as well as the volume of water it will contain at room temperature, 20°C, and this information is very helpful when testing liquids but is not needed when testing a solid by immersing it in water. First weigh the stone on a scale (A). Then completely fill the pycnometer with water, insert the stopper, and allow the excess to flow out. Dry the outside of the pycnometer and weigh it (B). Finally, remove the stopper, put the gemstone in the pycnometer and replace the stopper. This will cause a volume of water equal to the volume of the gemstone to flow out of the capillary opening. Dry the pycnometer and re-weigh it with the stone inside (C). From these three weights we can calculate the SG of the stone. This is most easily understood by considering the stone/water/pycnometer as a system. We have organized the system in two different states, one with the stone outside the flask that is filled just with water, the other with the stone inside the flask immersed in water (having displaced water from the flask). The total weight of the system with the stone in air is A + B and this weight will be greater than C, which is the total weight of the system when the stone is immersed. The difference between these total system weights will be the weight of the water that was displaced when the stone was immersed. We then see that (A + B) – C is the weight of the displaced water. The definition of SG is the weight of an object divided by the weight of an equal volume of water, so the formula for calculating SG of a solid using the pycnometer is:

$$\text{Specific Gravity (SG)} = \frac{\text{weight of gemstone in grams (A)}}{\text{weight of displaced water in grams (A+B-C)}}$$

If a red gemstone weighs 1.80g, the weight of the water-filled pycnometer is 40.0g, and the weight of the pycnometer with the stone immersed is 41.3g, then the calculation 1.80 ÷ (1.80 + 40.0 – 41.3) sets the SG as 3.60. The gemstone with an SG of 3.60 is spinel and the stone tested is a red spinel, although it may be natural or synthetic.

Figure 5.3 The pycnometer or specific gravity bottle. It is engraved on the outside with its weight when empty and the volume of liquid it will contain at room temperature, 20°C.

Heavy liquids

Heavy liquids, or specific gravity liquids, provide a quick means to differentiate between gemstones on the basis of specific gravity. It is accurate for small gemstones and the liquids are portable, but overall the method is less precise than hydrostatic weighing and is too expensive for use with larger specimens. As with all SG tests, it cannot be used with mounted stones. Heavy liquids typically yield only relative results, not wholly precise ones, unless a specialized liquid is made and its SG determined by pycnometer.

Archimedes' Principle states that a body immersed in a fluid experiences an upward force equal to the weight of the fluid it displaces, therefore when a gemstone has the same specific gravity as the liquid in which it is immersed it will neither sink nor float but remain freely suspended. This is the basis for the principles of heavy liquid testing: a stone will float in a liquid of higher density than itself; a stone will sink in a liquid of lower density than itself; and freely suspend in a liquid of the same density as itself.

Heavy liquids were first employed in gemmology as a means to identify cultured pearls, which were denser than naturals, in the 1920s. The criteria required for a heavy liquid are high density and the capability to be thoroughly mixed with another liquid of lower density for the production of more exact liquids of intermediate density. There are many liquids that have the necessary criteria, and at various times a greater number of liquids were used; however, at present there are three basic liquids: (1) di-iodomethane (CH_2I_2), also known as methylene iodide, with a specific gravity of 3.32; (2) bromoform ($CHBr_3$), with a specific gravity of 2.88; and (3) monobromonaphthalene ($C_{10}H_7Br$), also known as 1-bromonaphthalene, with a specific gravity of 1.49. The first two liquids can be mixed together to produce intermediate densities or they can be diluted with a liquid of lower density such as benzene (SG 0.88) or toluene (SG 0.87), but because benzene is a carcinogen and both benzene and toluene are inflammable the typical choice is monobromonaphthalene.

Clerici solution is a concentrated aqueous solution of the formate and malonate thallium salts. It was popular because it has a high SG (4.15) and can be readily diluted with distilled water. Its principal advantage was that it could be used to test the higher SG gemstones such as garnet, corundum, spinel and diamond; however, thallium salts are poisons and Clerici solution is corrosive and very dangerous as it can be absorbed through the skin and remain in the body. It is not permitted in gemmological classes and is only occasionally used in laboratory settings.

Heavy liquids may be purchased individually or in boxed sets as a series of glass tubes with ground glass stoppers. A typical set may contain:

3.32 – undiluted di-iodomethane

3.05 – diluted di-iodomethane – tourmaline suspends

2.88 – undiluted bromoform

2.65 – diluted bromoform – quartz suspends

Individuals also frequently mix liquids of: 3.18 – diluted di-iodomethane – fluorite suspends; and 2.71 – diluted bromoform – beryl suspends, rises or sinks slowly.

Some sources recommend that di-iodomethane be diluted with monobromonaphthalene, but a better choice would be to dilute it with bromoform. Mixing liquids that are closer in SG allows tighter control of the mixing and can produce precise suspending of the test stone or the indicator more easily. It is possible to use either, but using bromoform makes it a little easier to produce a precise SG and the mixture would be slightly more stable in long-term storage.

Heavy liquids are very handy when optical readings cannot be obtained such as with nephrite and jadeite. A liquid of 3.18 where nephrite (SG 2.90 to 3.02) floats and jadeite (SG 3.30 to 3.36) sinks can be made by using a small amount of di-iodomethane in a narrow cylindrical glass container with a wide mouth. Heavy liquids are expensive so this type of container allows a smaller amount of liquid to occupy sufficient depth to give proper vertical distance in the liquid with which to observe the rise or fall of the stone without wasting liquid. Dilute the di-iodomethane with the bromoform one drop at a time. Stir the mixture with a clean glass rod after every drop to be certain that it is mixed thoroughly and is uniformly dense. An indicator stone of transparent fluorite (SG 3.18) should just sink below the surface and suspend when the desired specific gravity has been obtained. Sometimes two indicator stones are used, one that floats and one that sinks, indicating that the liquid's specific gravity is between those two limits. If one is able to find a suitable indicator, two stones are unnecessary. Fluorite is a good indicator because it has a very consistent SG. Keep in mind that it is also possible to use synthetic stones as indicators as in most cases they have the same constants as their natural counterparts. An exception is emerald; synthetics may have SGs as low as 2.65 whereas the lowest natural is known to be 2.67. It is a good idea to use a crystal or irregular-shaped gemstone if possible so that it will not be confused with test stones.

Mix with care and avoid direct contact with the liquids. Leave the indicator stone in the fluid so that its position can be observed to ensure that the liquid remains accurate. The SG of the liquid can change over time due to one constituent evaporating faster than another, contamination or temperature change. Seal the glass jars tightly and keep them in a dark place, as all organic compounds, including heavy liquids where a halogen replaces the hydrogen in the molecule, will darken on exposure to sunlight. This is due to decomposition caused by the release of free iodine. Darkened liquids make it hard to see what is going on with the indicator and test stones, so it must be avoided. A few strips of copper in the bottle stop this as copper combines with the free iodine and prevents darkening. Specific gravity liquids are accurate for one temperature, room temperature in the testing process, so avoid excessive handling of the bottles as heat from the hands can warm the liquid and cause inaccuracy.

The testing procedure is simple. Clean the stone and tongs with alcohol and, when they are dry, hold the stone in the tongs and place it into the liquid releasing it below the surface. Observe the reaction. A stone that sinks rapidly has an SG much higher than the liquid and a stone that rises rapidly has an SG much lower than the liquid. If it suspends it is the same SG as the liquid, and if it sinks to the bottom or floats to the top slowly, its SG is near to that of the liquid but slightly higher or slightly lower respectively. The stone and the tongs must be thoroughly cleaned after testing, or before retesting in a different liquid, to prevent damage or contamination. For some stones, as in the nephrite/jadeite example, only one liquid need be used, but if the stone is a complete unknown and several liquids may be needed it is a good general practice to start testing in a liquid of higher SG than the suspected stone as it is easier to remove a test stone from the top of the liquid than the bottom. In reality very few gemstones suspend; the typical result is a relative SG between the liquid in which it sinks and the liquid in which it floats.

It is unwise to immerse porous stones, such as turquoise and opal, as they may be damaged by discolouration. If absolutely necessary, they should be rinsed clean and carefully dried. Some plastics soften in the liquids and the liquids are also known to damage the adhesion layer in composite stones. Stones that have surface-reaching fractures can also absorb fluid and become damaged. It is best to remove all stones as quickly as possible.

An additional specific gravity liquid is obtained by mixing ten level teaspoons of table salt in a tumbler of water. This produces a solution with a density near to 1.13, which is suitable for distinguishing between amber (SG 1.08) and the common amber substitute Bakelite (SG 1.26 to 1.28) and other plastic imitations. Amber floats on the salt solution and the most common imitations sink. It should be noted that this solution does not separate amber from natural copal resins or the synthetic resin polystyrene.

Precise liquids

In instances requiring more than an approximation of SG, a special liquid can be made by the same procedure as that used to mix the 3.18 liquid, in order to suspend the test stone. This liquid then has the same SG as the test stone and the specific gravity of the liquid can be determined by pycnometer. Alternatively, if two pure liquids of known SG and RI are used to create the precise liquid, values of the two liquids can be plotted on a straight-line graph. The RI of a mixture can then show the SG of the mixture from that straight-line graph.

The pycnometer is capable of considerable accuracy but is time-consuming. As mentioned previously in the water displacement discussion, the flask is generally engraved with its weight when empty and the volume of water it holds in cubic centimetres. If not engraved, the flask must be weighed when empty and again when filled with the heavy liquid whose density is to be determined, and the difference between the two weights will give the weight of the liquid. The weight of an equal volume of water is known if engraved on the flask; this is the equivalent in grams to its capacity in cubic centimetres. If the flask is not engraved, this can be obtained by weighing the flask filled with water and then subtracting the weight of the empty flask. Then, by dividing the weight of the water into the weight of the liquid, the density of the liquid is obtained:

$$\text{Specific Gravity (SG)} = \frac{\text{weight of the heavy liquid}}{\text{weight of the water}} \text{ or } \frac{X}{Y}$$

If a transparent green gemstone suspends in a heavy liquid its specific gravity can be determined as follows. Suppose the weight of the empty

DENSITY AND SPECIFIC GRAVITY

Table 5.1 Specific gravity

Listed are the average range or mean values for the specific gravity of various gemstones. Many gemstones show greater or considerable variation by colour and variety, and as a result of impurities.

Gemstone	Specific gravity
Amber	1.08
Jet	1.33
Opal	2.10
Feldspar – moonstone	2.57
Quartz	2.65
Aquamarine	2.68 to 2.71
Emerald	2.69 to 2.75
Turquoise	2.75
Nephrite	3.00
Tourmaline	3.07
Spodumene	3.18
Fluorite	3.18
Jadeite	3.34
Peridot	3.34
Zoisite – tanzanite	3.35
Sinhalite	3.48
Diamond	3.52
Sphene	3.53
Topaz	3.56
Spinel	3.60
Garnet – hessonite	3.65
Garnet – pyrope	3.65 to 3.78
Chrysoberyl	3.71
Garnet – demantoid	3.85
Corundum	4.00
Zircon – low type	4.10
Garnet – almandine	3.95 to 4.20
Zircon – high type	4.60 to 4.80
Strontium titanate	5.13
Cubic zirconia	5.65 to 5.95
Gadolinium gallium garnet (GGG)	7.05

pycnometer is 35g (A) and it holds 50 cu cm of water, which weighs 50g (Y). If the weight of the pycnometer filled with the heavy liquid is 202g (B), then the weight of the heavy liquid (X) is (B − A = X) 202 − 35 = 167g. Therefore, the specific gravity of the heavy liquid is 167 ÷ 50 = 3.34. The specific gravity of the heavy liquid is 3.34 and the gemstone with that specific gravity is a peridot.

Formula for the replacement of gemstones with different SGs

A practical point that follows from the discussion on differing specific gravities is that stones of different species, but of the same dimensions, will weigh different amounts. The general principle is that a stone weighing 1.00ct with a higher SG will take up a smaller space by volume because it is denser than a stone weighing 1.00ct of a lower SG. Hence, if a diamond of a given size is required to be replaced by a cubic zirconia, then the required weight of the cubic zirconia necessary to appear the same size as the diamond can be obtained by the formula:

$$\frac{\text{SG of stone required} \times \text{weight of stone to be replaced}}{\text{SG of stone to be replaced}} \text{ or } \frac{5.65 \times 1.20 = 6.78}{3.52} = 1.93$$

Thus, if the specific gravity of cubic zirconia is 5.65, the specific gravity of diamond is 3.52 and the weight of the diamond to be replaced is 1.20ct, the formula determines that a cubic zirconia of 1.93ct is required.

LESSON 6

Light

Without light there could be no world as we know it. No splendour of the sea or sky, no glorious colours of the flowers in the field and no beauty of fine gems; however, to the gemmologist, light holds an even more vital interest than illuminating the beauty of gemstones and allowing us to see that beauty – that is, the usefulness it possesses as a means for gem identification. Almost all gemmological tests are optical tests involving light. The refractometer, microscope, spectroscope, polariscope and dichroscope all use the effects of light to discriminate between stones. This lesson deals with the nature of light and how light behaves when it impinges upon a medium, travels through a medium, or travels from one medium to another. It concludes with optical properties and the optical phenomena seen in gemstones.

What is the nature of light? Over the course of history there have been many hypotheses. The first to be based on theoretical physics was that of Dutch scientist Christiaan Huygens who proposed a *wave theory* of light circa 1678, which held that light travels in all directions as a series of waves in a medium he called the luminiferous ether. For light to have a waveform it seemed essential for it to proceed through some sort of medium, meaning any substance through which light may pass, referring to solids, liquids or gases. When the light source is that of a candle, gas flame or electric light, the atmosphere may appear to be that medium, but if we consider the greatest natural producer of light, the sun, this can no longer be the case. The sun is approximately 150 million kilometres from the Earth, far beyond the Earth's atmosphere into outer space which, although not a perfect vacuum, has only a low density of particles; hence, the assumption of a hypothetical medium that is elastic and weightless, known as the ether.

Huygens's theory was challenged by that of Sir Isaac Newton who suggested that light was composed of corpuscles, infinitesimally small discrete bundles of energy now called photons, projected in all directions in straight lines from a source. He published this theory in 1704 in *Opticks* and his *corpuscular theory*, otherwise known as the *particle theory of light*, was the predominant belief of the 1700s.

In 1845, Michael Faraday discovered that magnetic field directions affect the plane of polarization of linearly polarized light. This led Faraday to propose that light was a high-frequency electromagnetic vibration, which could propagate even in the absence of a medium such as the ether. James Clerk Maxwell discovered that self-propagating electromagnetic waves travel through space at a constant speed – that speed being equal to the speed of light – and concluded that light was a form of electromagnetic radiation.

Today, it is known that light is in fact electromagnetic radiation and that its behaviour can be explained in terms of both wave theory and particle theory. This position, known as wave-particle duality, follows from the twentieth-century research of Max Planck, Albert Einstein and others, and holds that all energy, and by implication all matter, exhibits both wave-like properties and particle-like properties: all particles also have a wave nature and all waves also have a particle nature. It is a central concept in quantum electrodynamics, which mathematically describes how light and matter interact in terms of phenomena involving electrically charged particles interacting by means of photon exchange. Although both wave and particle theories are equally valid, to understand most optics and optical phenomena in gemmology, it is generally sufficient to approach the subject of light from the wave theory perspective.

The electromagnetic spectrum and light waves

The electromagnetic spectrum refers to the range of possible wavelengths of electromagnetic radiation. All light is a form of electromagnetic radiation, and visible light refers to those wavelengths that can be captured by the visual system – the eyes, the visual pathways and the visual centres of the brain – to generate visual perception, the sensation we commonly call 'sight'. Visible light constitutes a very small range of wavelengths near the centre of the electromagnetic

LIGHT

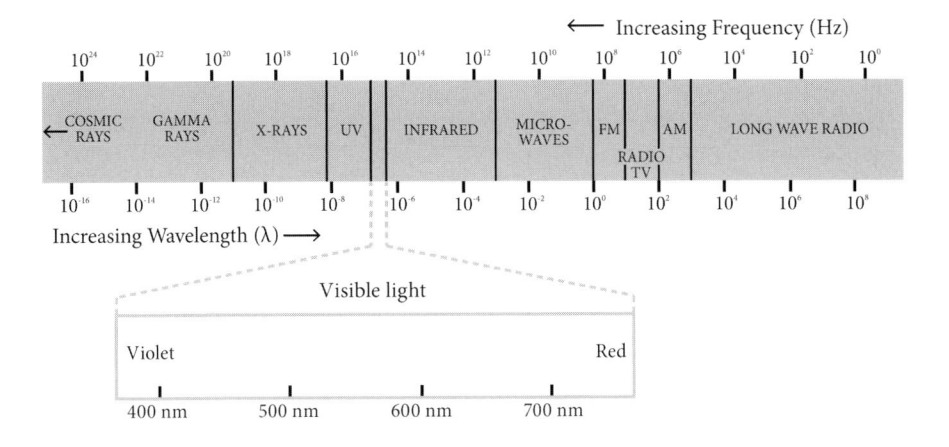

Figure 6.1 Electromagnetic spectrum.

spectrum. White light, also known as the visible spectrum, is composed of a mixture of red, orange, yellow, green, blue and violet light, each of which has a different wavelength. Therefore, the colour of light varies with and depends upon wavelength. Red has the longest wavelength, and this progressively diminishes through orange, yellow, green, blue and violet.

The lengths of light waves are so small that measurement by ordinary standards, such as metres, is impractical. For instance, one of the wavelengths of yellow light is that known as the D_1 line of sodium. It has a wavelength of 0.0000005896m. Wavelength becomes more understandable if it is expressed in nanometres (nm). A nanometre is one thousand-millionth (or one billionth) part of a metre. This makes the wavelength of the D_1 line of sodium 589.6nm. For a time it was convention to refer to wavelength in Angstroms, a unit commonly referenced in older texts, and conversion between the two units is simple as there are 10 Angstroms in 1nm. The wavelength of the D_1 line of sodium is 5896 Angstroms.

When energy has a shorter wavelength than violet, it is invisible light known as ultraviolet light. Shorter than ultraviolet light are those radiations discovered by Röntgen in 1895, commonly known as X-radiation or X-rays. Shorter still is gamma radiation emitted by radium and the shortest of all is cosmic radiation. At the other end of the visible spectrum is invisible radiation of longer wavelength than red, termed infrared radiation. This is heat-radiation. Beyond this are

Table 6.1 Visible light

Wavelength (in nm)	Colour
<400nm	Ultraviolet
400–430	Violet
430–490	Blue
490–510	Blue–green
510–550	Green
550–575	Yellow–green
575-590	Yellow
590–630	Orange
630–650	Orange–red
650–700	Red
700–750	Deep red
>750nm	Infrared

microwaves, and longer still are radio and TV waves. The longest of all are those of long wave radio. It is worth noting here that it is commonplace to discuss light and refer to radiations in terms of rays – for example, gamma rays, X-rays, cosmic rays and a ray of light. Rays are imaginary lines in the direction of radiant flow; they are simply an invented means to visualize the concept of how visible light, or other electromagnetic radiation, interacts with media. They are not real in the sense that the waves and particles of light are real.

When waves are mentioned what most often comes to mind is the surface waves created when a stone is thrown into the water. But unlike ripples on the surface of water, a light wave travels in straight lines, but it undulates, meaning it moves back and forth at right angles to the direction of travel. These wave vibrations can occur in all planes at right angles to the direction of propagation, meaning the direction in which the light is travelling.

The speed of light in a vacuum is a universal constant: 300 000km/sec. Light is transmitted slower through other media. The amount the speed of light is slowed depends upon the composition and structure of the medium through which it is transmitted and the vibration direction and wavelength of the light. In gemmological studies it is sufficient to know that light slows proportionately as the optical density of the transmitting medium increases.

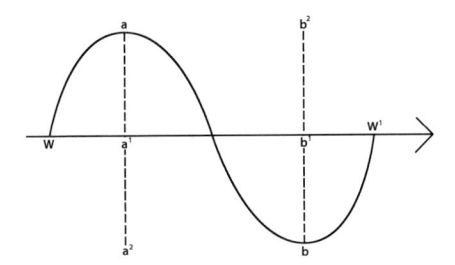

Figure 6.2 Wave form illustration: the direction of propagation is indicated by the arrow. Vibration direction is the direction between a and a_2 and b and b_2. Wavelength is equal to the distance W to W^1. Amplitude is the distance a_1 to a and b_1 to b.

Light waves are described by their direction of travel and their vibration direction as well as by other properties. Amplitude refers to the magnitude or strength of the lateral vibration motion of the wave. Amplitude determines the intensity of the energy. In the visible light range this is perceived as relative brightness: the greater the amplitude, the brighter the light. Frequency refers to the number of wave cycles that pass a fixed point within a designated time. Frequency is established when the radiation is generated and is invariable. In simple transmission, the frequency remains unchanged even when the velocity changes. For frequency to change, energy must be either subtracted from or added to the light as in Raman shift or luminescence. Wavelength is measured as the linear distance between two consecutive points in identical positions on a wave, meaning having identical amplitudes and vibration directions. In optics the importance of wavelength is that it gives information about frequency, although in gemmological studies wavelength is the property referred to most often. The speed of light, frequency and wavelength are related by the equation: $V = fw$. Because frequency is invariable, wavelength has a direct relationship to velocity: the higher the velocity, the longer the wavelength. And wavelength has an inverse relationship to frequency: the greater the wavelength, the lower the frequency.

Reflection

When a ray of light strikes a front-surfaced plane mirror at an oblique angle the ray is turned back, or reflected away from the surface in the opposite direction to which it arrives and at a similar angle. This may be best explained by example. In Figure 6.3, PM is a front-surfaced plane mirror. IO is a ray of light, the *incident* ray, falling on the mirror at O. O is called the origin or the point of incidence. The ray of light is reflected from the surface along OR which is the reflected ray. NO is termed the *normal* at the point of incidence. The normal in optics is an imaginary line perpendicular to the surface at O where the light ray strikes the surface. It is the base line from which all angles in optics are measured. The angle ION is the angle of incidence (i) and it is equal to NOR, the angle of reflection (r).

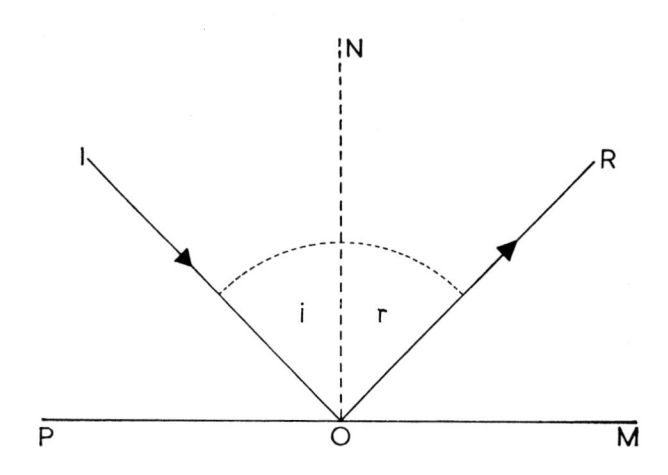

Figure 6.3 Reflection from a front-surfaced plane mirror.

The laws of reflection

1. The angle of incidence is equal to the angle of reflection.
2. The incident ray, the normal and the reflected ray all lie in the same plane.

Other than colour, reflection, sometimes in combination with other effects, is responsible for most features of beauty seen in cut gemstones including: lustre and sheen, and such effects as asterism (star-stones) and chatoyancy (cat's eyes). Reflection is strictly a surface effect; none of the reflected light passes through the surface from which the reflection occurs. The reflecting surface may, however, be entirely enclosed within a gemstone, so light travels through the surface of the gemstone and then reflects from the surface of the inclusion. Reflections that yield sheen occur at surfaces inside the stone, so are termed *near-surface reflections* in relation to the surface of the gemstone.

Refraction

When light falls obliquely on the surface of a transparent medium such as a gemstone, some of the light is reflected at the surface at an angle equal to the angle of incidence, and some of the light will enter the stone where it slows down and changes path. The ability of a medium to influence the directional travel of light is termed *refringence* or *optical density*. The measure of optical density is refractive index (RI). Because the gemstone is an optically denser medium than the air, instead of travelling through in a straight line, the light changes speed and alters course. Refraction is defined as the change of direction of light due to a change in velocity when it passes between media of differing optical density. Refraction is responsible for the common observation that a straw leaning against the side of a glass of water appears to bend where it enters the water; this is because the light bends as it enters or exits the water, an optically denser medium than the air. If it were possible for the straw to come out of the bottom of the glass and back into the air, it would appear to bend again, going parallel to its original path.

For refraction to occur in an isotropic medium, meaning a medium having identical optical properties in every direction, it is necessary for the light to strike the surface of an optically denser or rarer medium at an oblique angle. Light that strikes the surface at 90° to the interface, along the normal, slows down or speeds up but continues in a straight path. At this orientation, the light does not change path because each end of the wave train strikes the medium at exactly the same time. When a wave train strikes at an oblique angle, one side of the wave train enters the medium and slows down or speeds up first, while the other

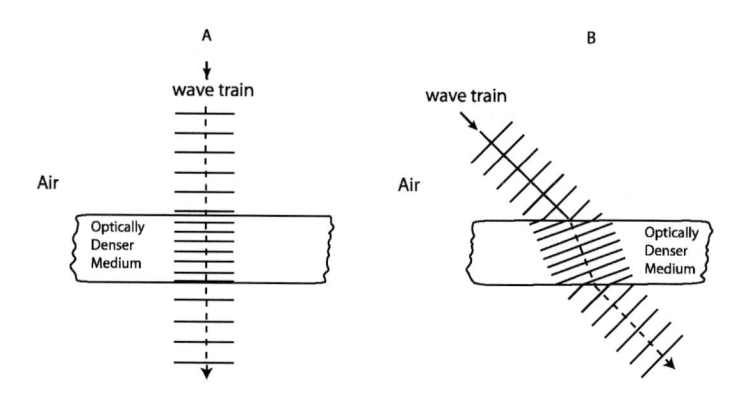

Figure 6.4 Wave train entering a denser medium. In diagram A, both ends of the wave train enter the medium at the same time; the wave train slows but it does not change path. In diagram B, the left side of the wave train enters first and slows while the right side continues at its original speed; this pulls the wave train towards the normal. Upon exiting, it resumes its original path.

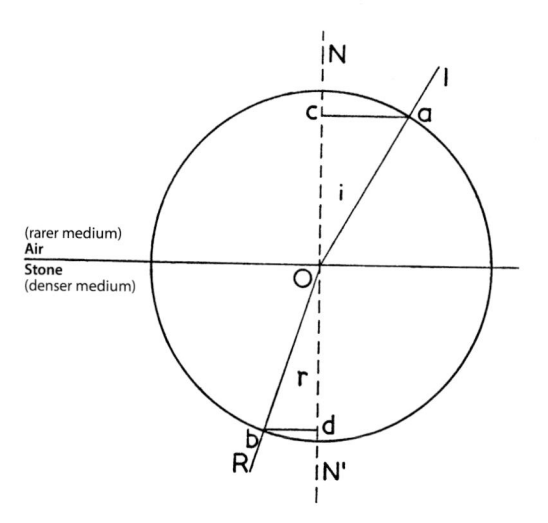

Figure 6.5 Refraction of light. A circle drawn with the centre at the point of incidence of the light, and with any radius that cuts the incident and refracted rays at **a** and **b** respectively, will, when perpendiculars **ac** and **bd** are drawn to the normal, produce right-angled triangles. Therefore the ratio of the side **ac** and the hypotenuse **aO** gives the sine of the angle of incidence, and the ratio of the side **bd** and the hypotenuse **bO** gives the sine of the angle of refraction.

end is still in the air travelling at its original velocity. This difference in speed pulls the wave train and it changes course.

In Figure 6.5, IO is the incident ray, OR is the refracted ray and NN^1 is the normal at the point of incidence. It will be seen at once that the angle of incidence (i), ION, is greater than the angle of refraction (r), RON^1. Any increase or decrease in the angle of incidence causes a proportionate increase or decrease in the angle of refraction. Dutch astronomer and mathematician Willebrord Snellius (Snell) discovered that for isotropic media there is a constant ratio between the sines of the angles of incidence and refraction for any two media in optical contact.

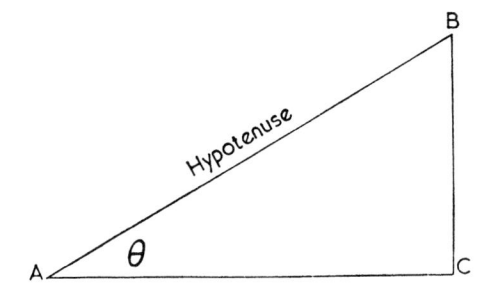

Figure 6.6 In a right-angled triangle the sine of an angle is the ratio of the length of the side opposite that angle to the hypotenuse, the side opposite the right angle. In this example:

$$\frac{BC}{AB} = \text{Sine } \theta$$

Snell's laws of refraction

1 The sine of the angle of incidence bears to the sine of the angle of refraction a definite ratio that depends only upon the two media in contact and the nature of the light.

$$\frac{\text{Sine of angle of incidence}}{\text{Sine of angle of refraction}} = \text{a constant}$$

2 The incident ray, the normal and the refracted ray all lie in the same plane.

Two principles to understand are: (1) when light travels from a rarer to a denser medium it is refracted towards the normal. This is the case with light travelling from air to a gemstone. In this case the angle of refraction is smaller than the angle of incidence; and (2) when light travels from a denser to a rarer medium it is refracted away from the normal. In this case the angle of refraction is greater than the angle of incidence. When light travels from air to another medium, the ratio between the sine of the angle of incidence and the sine of the angle of refraction is called the *refractive index* (RI) of the medium. Refractive index, a measure of optical density, is a valuable discriminatory tool for the gemmologist. Refractive index and its measurement is the subject of Lesson 7.

Dispersion

When a ray of white light is passed through a transparent medium with parallel sides, such as a glass plate, the emergent ray is found to be parallel to the incident ray although often laterally displaced, as in Figure 6.7. However, when a ray of white light is passed through a transparent medium with two inclined faces, such as a glass prism, the white light is resolved into the familiar colours of the rainbow. Dispersion occurs as white light is sorted by wavelength to yield rainbow colours when it passes through inclined faces of a transparent medium denser than air. Dispersion is caused by simple refraction since refractive index varies with the wavelength of light and not all colours are refracted to the same degree. The bending of light is inversely proportional to the length of the wave: the longer the wavelength, the less it is slowed down and bent by the denser medium, therefore the red end of the spectrum is bent the least and the violet end is bent the most.

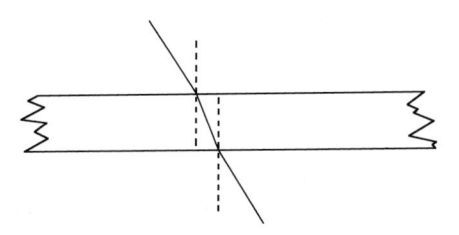

Figure 6.7 Path of light rays through a parallel-sided glass plate.

LIGHT

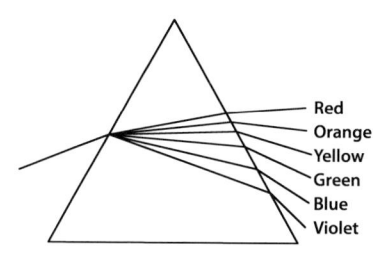

Figure 6.8 Dispersion: the separation of white light into a spectrum of rainbow colours. The prism analyses the white light into its constituent rainbow colours and arranges them according to their wavelength. The longer wavelengths, red light, are bent the least while the shorter wavelengths, violet light, are bent the most.

Dispersion is a measure of the difference in RIs for two different wavelengths. Theoretically, dispersion should be measured at the extremes of the violet and red ends of the spectrum, but in gemmological practice it is measured at two fixed wavelengths: the B and G Fraunhofer lines in the sun's absorption spectrum. The B line is at 686.7nm in the red and the G line is at 430.8nm in the blue-violet. For example, the refractive index for the G line in diamond is 2.451 and the refractive index for the B line is 2.407, producing a difference of 0.044 as the dispersion value. The length of the path of light through the stone has no bearing on the amount of dispersion produced, meaning large diamonds have no greater power of dispersion than small diamonds. Dispersion depends on the angle of incidence, the angle between the two inclined sides and the B–G value. Dispersion occurs at the moment light passes through the surface as it enters or exits a dispersive medium.

Dispersion is responsible for the fire in gemstones and is strongest in stones such as diamond, and its imitations – rutile and strontium titanate – as well as demantoid garnet, sphene and zircon. All transparent gemstones produce dispersion, it is a property of minerals, but those that are colourless allow it to be seen to a more appreciable extent. In reality, it is very seldom measured, although when estimated from the appearance of a stone it is useful for gem identification in certain instances. The characteristic dispersion of diamond becomes quite recognizable with experience.

Table 6.2 Dispersion

Listed are the average or mean values for the dispersion of various gemstones.

Rutile	0.28 to 0.30
Strontium titanate	0.190
Cubic zirconia	0.060
Demantoid garnet	0.057
Sphene	0.051
Diamond	0.044
Zircon	0.039
Garnet–most	0.027
Garnet–pyrope	0.022
Spinel	0.020
Peridot	0.020
Corundum	0.018
Tourmaline	0.017
Chrysoberyl	0.015
Topaz	0.014
Beryl	0.014
Quartz	0.013

Interference

Interference is a physical phenomenon due to the wave properties of light and is caused by waves travelling in the same medium. If a ray of light impinges on a thin film, such as the surface of a soap bubble, interference results. In Figure 6.9, AB is the light incident on the surface, and part of the ray is reflected along BC and part is refracted into the medium along BD. In turn, BD is reflected along DE and then finally refracted out again along EF. BC and EF are parallel and close together, but EF will be retarded owing to its longer passage along BDE. Hence, if the incident light is white, some of the waves will be out of phase because they are split and then recombined after some have travelled a short distance more. When waves are out of phase, meaning with the peak of one wave coinciding with the trough of another, and they then combine, it causes cancellation or nullification. When waves are in phase, meaning with the peaks and troughs coinciding with one another, and they then combine, the amplitude of the wave will be increased as the two are in effect added together. Interference causes nullification, amplification or any intermediate effect between these extremes, and it results in certain colours being intensified or removed from the incident white light. The greater the path difference, the more the waves will be out of phase. The thicker the film and the more oblique the angle of the incident ray, the greater the path difference will be. Different colours are intensified with different thickness and different angles of incidence. These colours will differ according to the angle at which the surface is viewed.

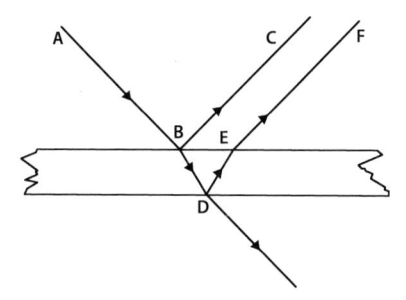

Figure 6.9 Interference of light.

Thin film interference is often just termed 'interference', but colour may also be produced by interference due to diffraction in which light waves passing an edge between media with two different RIs, or between transparent and non-transparent media, become out of phase. In minerals or artificial structures such as a diffraction grating, where enough of these edges are present and spaced evenly, it causes interference. Both phase interference and interference by diffraction may be the cause of colour in pearl, opal and other gem materials. Interference by diffraction is often just termed 'diffraction'.

Transparency

Diaphaneity is the mineralogical term that refers to the degree of transparency: the freedom with which light is transmitted through a mineral. Transparency affects a gem's beauty and value. In order to determine transparency a gemstone should be polished because rough gems commonly have frosted or coated surfaces that adversely affect the passage of light. Transparency is classified in degrees as follows: transparent, semi-transparent, translucent, semi-translucent and opaque. In *transparent* gemstones an object viewed through the stone can be seen clearly with a distinct outline. If there were printing beneath the stone it would be possible to read it – in theory. In practice, it is not possible to see through a well-proportioned brilliant cut diamond because light entering the stone is reflected back to the observer; however, if diamond or any other transparent stone were cut like a pane of glass, one could see through it and make fine discriminations. Gemstone examples include: diamond, spinel, emerald, quartz and topaz. In *semi-transparent* gemstones the object would be visible, but blurred with no distinct outline. Although considerable light penetrates the stone, if there were printing beneath the stone, it would be possible to tell there were words, but it would not be possible to read them. Gemstone examples include amber and chalcedony. In *translucent* gemstones some light passes through but no objects can be seen – for example, as in opal and jadeite. In *semi-translucent* gemstones light is only transmitted through at thin edges – for example, as in turquoise. And in *opaque* gemstones no light passes through – for example, as in lapis lazuli, malachite, pyrite and jasper.

The above descriptions refer to specimens of ordinary thickness because even opaque minerals are translucent if ground to thin sections. The thicker the stone the more the passage of light is restricted; therefore, stones should be neither excessively thick nor extremely thin when judging transparency. Transparency is also affected by colour since more deeply coloured stones transmit less light than paler-coloured stones. As well, feathers, included crystals and fibrous inclusions also obstruct the transmission of light. Overall, diamond is the most transparent gem, but individual stones differ in their degree of transparency from transparent to translucent, because a profusion of microscopic inclusions can make a stone cloudy, thus reducing its ability to transmit light.

Lustre

Lustre is produced by the reflection of light from the gemstone. The type of lustre is dependent on the structure of the mineral. Opaque minerals such as pyrite, and opaque metals such as gold, have metallic lustre. This is a highly reflective lustre resulting exclusively from surface reflection. A broad range of terms, typically with no strict boundaries, describes the

Table 6.3 Lustre

Adamantine	The highest degree of lustre. Only seen in diamond.
Sub-adamantine	A lustre less brilliant than adamantine but still very high. Typically transparent to translucent stones with high refractive index such as zircon and demantoid garnet.
Vitreous	A glass-like lustre. Most often occurs in transparent to translucent minerals with moderate to low refractive index. This is the most common lustre of gemstones: corundum, topaz, quartz, beryl, tourmaline and many others.
Resinous	The appearance of a smooth-surfaced plastic. Characteristic of amber.
Waxy	The wax-like lustre characteristic of jadeite, chalcedony and turquoise.
Pearly	Characteristic of pearl.
Greasy	Resembling fat or grease. Often occurs in minerals having a profusion of sub-microscopic inclusions. Typical of soapstone.
Silky	A fibrous lustre typical of minerals with a parallel arrangement of extremely fine fibres such as satin-spar.
Metallic	Very high lustre exhibited by opaque minerals such as marcasite, hematite and pyrite, and opaque metals such as gold and platinum.

lustre of all other, non-metallic, minerals. Furthermore, the lustre of individual specimens can vary within a species. For these reasons it is possible to find different terms attributed to the same mineral in different sources. In some cases lustre is described by a combination of two terms: waxy–greasy. Although lustre is defined in gemmology as the surface brilliancy of a gem in reflected light, certain types of non-metallic lustre – for example, pearly and silky – are produced by surface reflection in combination with some slight contribution from sub-surface reflection.

Several factors affect the lustre of a gemstone. There is a direct relationship between RI and lustre: high RI produces high lustre – such is the case with diamond. Polish also affects lustre because the flatter or more perfect the surface, the higher the lustre; however, polish is not capable of increasing lustre beyond the natural reflecting power of the gemstone itself. Lustre is generally higher in harder stones because harder stones take a better polish, although there is no direct relationship. As a general rule, transparent materials are not highly reflective; the most reflective are typically opaque, but diamond and other gemstones are exceptions. Lustre is adversely affected by surface-adhering dirt and grease.

Sheen

In contrast to lustre, which is due to surface reflection, sheen is due to near-surface or sub-surface reflection of light. It is a general umbrella term for a variety of optical phenomena including: chatoyancy, asterism, iridescence, orient, play of colour, aventurescence and adularescence. *Schiller*, from the German for 'twinkle' or 'glitter', is observed as a sparkle in the sheen that may be the result of the mechanism of the sheen itself or the mechanism of the sheen in combination with irregularity of the surface structure. It is observed on unpolished surfaces of sunstone or moonstone, but should not be used to refer to any sheen seen on a polished gemstone as it has been incorrectly used to describe various effects; these phenomena are better described by other terms.

Chatoyancy

The cat's eye effect, known as chatoyancy, is an optical effect in which a single mobile streak of light is reflected from parallel hair-like inclusions, fibres or channels within the stone. The band of light is at right angles to the direction of the inclusions and may be likened to the band of light seen when shining a point source of light on a spool of silk thread. The inclusions may be coarse or fine. The bright streak is mobile and moves across the stone when the stone, light source or observer moves. The effect is created by a series of flecks of light from successive inclusions. Chatoyant stones are typically cut as cabochons, meaning with a dome-shaped top, with the base of the cabochon parallel to the plane of the inclusions. In this orientation the curved surface confines the sheen to a single bright streak of light. Good-quality chatoyant stones have a single, centred eye and in the finest stones the eye is thinner and sharper. The rarest and most expensive cat's eye is the cymophane variety of chrysoberyl in which the band of light is due to hollow tubes. The term cat's eye without a qualifier should only be used to refer to chrysoberyl. The quartz cat's eye is not as sharp because the fibres are coarser. In crocidolite, or tiger's eye, the channels are the fossilized remains of asbestos fibres replaced by chalcedony.

Asterism

Asterism is a multiple chatoyancy effect caused by two or more sets of parallel inclusions which produce crossed bands of light on top of the stone in a star shape, and hence the name star stones. The inclusions lie parallel to the horizontal axes and at right angles to the vertical axes, and to show the effect the stone must be cut as a cabochon with the vertical axis upright. Fine-quality stones have a centred star, good colour and crisp rays that are not doubled. Star ruby and star sapphire are the finest varieties. Corundum belongs to the trigonal system and when the rutile needle inclusions run in three directions parallel to the faces of the first-order prism, the bands of light cross at 120° to produce a 6-rayed star. If there is a second set of inclusions parallel to the faces of the second-order prism, the bands of light cross at 60° – producing a 12-rayed star. Garnet, spinel and diopside may also show stars, but in this case with 4 rays. Garnet occasionally displays 6 rays. With all these stones the star is visible in *reflected* light and the proper term for this effect is *epiasterism*. Rose quartz may also show a star effect in *transmitted* light and this effect is termed *diasterism*.

Iridescence and labradorescence

Iridescence is a thin-film interference phenomenon seen on surfaces in which colour changes as the angle of viewing changes. It is familiar to us as the rainbow-coloured light effect seen on soap bubbles and oil slicks. It results from multiple reflections from multi-layered semi-transparent surfaces in which interference of these reflections intensifies or nullifies certain wavelengths, thereby reinforcing some colours and removing others. It is characteristic of iris quartz, rainbow obsidian and fire agate. The iridescent sheen seen on labradorite is usually referred to specifically as labradorescence. Different samples may show different colours; it can show any colour, but in any particular area it only shows one colour. Thin flakes of feldspar in the surface layer, a result of lamellar twinning, are the cause of the interference. The light effects seen on labradorite are actually a combination of two sheens: iridescence and aventurescence. Iridescence is a broader term than labradorescence and is usually associated with pure spectral colours or metallics. It is a general term for interference by reflection.

Orient

Orient is the subtle rainbow colour phenomenon seen on fine pearls and mother-of-pearl shell. Pearl nacre is composed of microscopic crystals of calcium carbonate that are often grouped together in 'tiles' that can vary significantly in size, thickness and distribution pattern from one pearl to another. Traditionally, orient has been explained as an interference effect due to the thickness of individual crystals, but more recent investigations indicate it is a diffraction phenomenon. Not all pearls display orient as it is only possible to create the effect in those whose structure is appropriate to produce the phenomenon. As a diffraction phenomenon the colours seen would be due to, and depend upon, the size, pattern and distribution of tiles of calcium carbonate in the form of aragonite across the surface of the pearl. The brightness of the resulting colours depends on how uniform the microstructure of aragonite tiles is as this determines how effective it will be as a two-dimensional diffraction grating. With this explanation, it would not be proper to refer to orient as an iridescent effect as it is not the result of thin-film interference.

Play of colour

Precious opal's play of colour is a rainbow-coloured effect in which the colours seen at each location progress through a partial series of hues when the stone, light source or the observer moves. Once thought to be due to interference, it is now known to be due to Bragg diffraction. The opal structure is a regular stacking of spheres of amorphous silica with regular voids in between. This forms a 3-D cubic lattice that produces a 3-D diffraction grating. Bragg diffraction causes pure spectral colours to be seen at different angles. Opal is structured in patches containing uniformly sized spheres, with the size varying from patch to patch. The size of the spheres determines the longest wavelength that can be diffracted. Small spheres only diffract violet and blue, the shorter wavelengths, and progressively larger spheres diffract green, yellow, orange and red as the wavelengths and sphere sizes increase. Some sources state that large spheres that produce red can also produce all of the other colours of the rainbow, but this is incorrect. Spheres of the same size diffract a range of colours, but not the full spectrum. Appropriately sized spheres can produce: red, orange, yellow, green – or, if larger, just red, orange and yellow; orange, yellow, green and blue; or yellow, green, blue and violet – or, if smaller, just green, blue and violet. The better the quality of the opal, the greater the red flash. The rarest and most valuable is black opal with an evenly distributed red flash. Play of colour should not be termed 'fire' because in gemmology the rainbow colours caused by dispersion are called fire and dispersion is not an issue with precious opal. Further complications arise because of a variety of opal known as 'fire opal'; this is named after its body colour and shows no play of colour in most samples.

Aventurescence

Aventurescence is a reflection effect similar to spangled glitter. It is caused by light reflections from minute parallel flat surfaces which produce tiny dispersed twinkles of reflected light. The spangles sometimes appear to some extent metallic or iridescent. The names of materials showing the effect are derived from the terms aventurine feldspar (sunstone), aventurine quartz (called simply aventurine) and aventurine glass (known frequently by the misnomer goldstone).

Adularescence

Adularescence is the term used for the blue or white sheen of moonstone. The effect is due to the unique structure of the gem: extremely tiny inclusions of albite in orthoclase feldspar. It is often reported that the moonstone structure is alternating thin layers and that the effect is due to interference at thin films; however, this is wrong. If the effect were due to interference there would be some instances of other colours as in labradorite, but moonstone is always a white or blue sheen.

When moonstone crystallizes in nature it is at a higher temperature and pressure and the sodium and potassium are distributed through the aluminium silicate in a structure that is stable at that higher temperature/pressure. As the crystal slowly cools, the smaller amounts of sodium become unstable in their location in the structure at the lower temperature. The sodium migrates through the structure, exchanging places with the potassium and groups together with other sodium aluminium silicate to form sub-microscopic crystals of albite as inclusions within the host orthoclase. Thus the composition of moonstone is orthoclase feldspar with innumerable tiny inclusions of albite uniformly scattered throughout. As the tiny albite crystals grew in place inside the host orthoclase the surrounding structure of the orthoclase imposed enormous influence on the orientation of each new albite crystal so that every one of these tiny crystals came to have exactly the same orientation within the host. The sheen seen in moonstone is light reflecting off the parallel surfaces of the albite inclusions. Because each inclusion is so small it is not seen as spangles, but rather is blended together as a uniform sheen, which is white in most moonstone.

The cause of the blue colour seen in fine-quality moonstone is somewhat similar to the scattering phenomenon that makes the sky blue, but is distinctive as a separate phenomenon because in moonstone the blue is seen in transmitted and reflected light. With the sky, the blue colour is in the scattered light and transmitted light gives the red colour of sunrise and sunset. When particles that cause scattering are smaller than the wavelengths of light, then shorter wavelengths are more strongly scattered. This is what gives blue sky. The mechanism is Tyndall scattering and the colour is known as Tyndall blue. The smaller and more numerous the particles, the stronger the Tyndall blue.

When the particles that cause the scattering are larger than the wavelength of light but in the same order of magnitude, the scattering is more complex. Gerhardt Mie developed a mathematical description explaining the relationship between particle size and direction of scattering related to wavelength. For particles that are about twice the size of the wavelength of light, Mie found that red and orange wavelengths are scattered diagonally sideways, leaving more blue wavelengths in the forward-travelling light. It is this Mie scattering that gives fine moonstone the blue colour in its sheen. When the particles that cause scattering are random-sized but mostly significantly larger than the wavelength of light, then all wavelengths are randomly scattered and/or reflected and the effect seen is that of opalescence.

Opalescence

Opalescence should not be confused with the spectral colours seen in precious opal, an effect known as play of colour. Opalescence is the milky appearance seen in certain specimens of moonstone and common opal. It is an optical effect due to internal reflection and the scattering of light from many tiny inclusions resulting in a type of cloud, similar to that seen when dust particles in a room are illuminated by a beam of light. The cause is scattering by random-sized particles, usually larger than the incident wavelengths.

Brilliance

Brilliance is the brightness of reflections from within a gemstone seen in contrast to adjacent or surrounding darkness. In simple terms, it refers to the noticeability of reflections of internal light reaching the eye from a gemstone in the face-up position and the visual appeal of bright and dark contrasting patches within the stone. Brilliance is sometimes used loosely to refer to life, but brilliance is a static light effect and life is a dynamic or changing combination of light effects. Diamond has a small critical angle of 24°and can display a higher degree of brilliance over a greater range of viewing angles than other transparent stones, and its characteristic brilliance is described as hard or sharp.

Brilliance is affected by: (1) polish – poor polish impairs reflection of light; (2) refractive index and critical angle – the higher the RI, the smaller the critical angle; and the smaller the critical angle, the greater the potential for brilliance over a larger range of viewing angles; (3) transparency – any obstacles, inclusions or phenomena affecting transparency affect transmission of light; (4) cut proportions – poor proportions allow for loss of brilliance because of blockage of incident light by the head of the observer or by light leakage; (5) symmetry – greater symmetry yields more large-flash fire and scintillation while even a slight reduction in symmetry breaks up the reflections into smaller patches; and (6) cleanliness – any grease or film adhering to the surface lowers the differential between the RI of the gemstone and the RI of the surrounding medium, which is presumed to be air. In effect, it lowers RI, which in turn widens critical angle, reducing brilliance.

Fire

The resulting spectral colours produced by dispersion are commonly called fire. When the spectral colours are observed in a gemstone, you are not seeing dispersion, but are seeing fire produced by dispersion. The dispersion of a gemstone is a constant and is always the same. Diamond always has a dispersion of 0.044, but that is not to say all diamonds exhibit the same degree of fire. The degree of fire produced in any one stone is influenced by several factors: (1) simultaneous bright/dark contrast – most obviously controlled by the size of the light source. Size is more important than type: a point source is best. Fluorescent light is diffused because of its large physical size and is a very poor source for showing fire; (2) distance of the light source – the farther away the light source, the purer the colours; (3) distance of the observer – the farther away the observer, the purer the colours; (4) size of the facet – the smaller the facet, the purer the colours; (5) brightness – brighter light improves fire by increasing simultaneous contrast and causing the eye opening to become smaller, resulting in fewer but purer colours. Lights as bright as possible are best; and (6) angle(s) of incidence – refers to cut. Pavilion proportions are far more important than crown proportions.

It was stated previously that the distance light travels through a stone has no bearing on the degree of dispersion, but there is still a difference in the fire exhibited by small and big stones. Purer colours are seen in smaller stones at arm's length, but fire can be seen at greater distances from bigger stones. A greater distance yields pure spectral colours in large stones, but this same distance may make it impossible to see the fire from smaller stones.

Scintillation

Scintillation refers to flashes of light seen as a sparkle effect, due to motion caused by movement of the observer, the light source or the gemstone. It is a dynamic light effect caused by total internal reflection. Factors affecting scintillation include: (1) number of facets – each facet reflects light to the eye of the observer as an individual flash. In general, the more facets, the greater the number of flashes; however, optimum facet number is relative to stone size; (2) quality of polish – the more highly polished the surface, the less diffused the resultant light and the stronger the scintillation; (3) facet angles and proportions – proper proportions ensure a maximum number of distinguishable reflections are seen with good simultaneous contrast; and (4) symmetry – misaligned facets cause the reflections to be broken up into smaller and less significant flashes. The best scintillation is displayed when the flashes are even and balanced throughout the face-up appearance so that the pattern is altered by movement of the observer, light source or gemstone in a similar way in all parts of the gemstone. A small, bright point source of light is the best for displaying scintillation; sunless daylight and fluorescent light are less adequate.

Life

Life is a combination light effect referring to brilliance, fire and scintillation. Obviously, the factors affecting each of these individual light effects are also the factors that affect life. If a gemstone is well cut, the majority of life is supplied by total internal reflection and a smaller percentage by surface reflection. In one case it can be an aid to identification with experience. A well-proportioned diamond in optimum lighting conditions has a distinctive appearance; the proportion of white light to coloured light and the interactions of brilliance, fire and scintillation are unique to diamond.

LESSON 7

Refractive Index and Measurement of Refractive Index

In Lesson 6 it was stated that when light falls obliquely on the surface of a transparent medium such as a gemstone, some of the light is reflected at the surface and some of the light is refracted into the stone where it slows down and changes path. Refraction is defined as the change of direction of light due to a change in velocity when it passes between media of differing *optical density*, or differing *refringence*. The phrase 'optical density' is more often used than the word 'refringence'; however, familiarity with the term 'refringence' can enhance understanding of 'birefringence' when we get beyond discussing isotropic materials. The change in direction of the light is refraction and the change in velocity is the cause of the refraction. Snell discovered there is a constant ratio between the sines of the angles of incidence and refraction for any two media in optical contact.

Refractive index

When light travels from air to another medium, the ratio between the sine of the angle of incidence and the sine of the angle of refraction is the refractive index (RI) of the medium. Refractive index is a measure of optical density that expresses on a relative basis the ability of a medium to slow down (and bend) light: the higher the RI, the greater the ability to slow down (and bend) light. Refractive index is different for different media but it is a constant for any given medium. The RI of olive oil differs from that of Vaseline but the RI of all olive oil is the same and the RI of all Vaseline is the same. Hence, this constant is a valuable factor for the discrimination of materials such as gemstones.

In the lesson on specific gravity, we saw that water is used as the standard or unity against which other relative measurements are taken. The standard used for refractive index is air. Strictly speaking, this should be a vacuum, where the speed of light is 300 000km/sec. The passage of light is slower in all other media because light slows down when it encounters electrons – it has to make a 'detour' around them; however, air is only slightly more optically dense than a vacuum and slows light by just 75km/sec, giving air the RI of 1.000294. This difference is negligible, so for gemmological purposes the refractive index of air is taken as 1.000, which is accurate to three decimal places.

$$\frac{\text{Sine of angle of incidence}}{\text{Sine of angle of refraction}} = \frac{\text{Speed of light in air}}{\text{Speed of light in the gemstone}}$$

Optical density must not be confused with physical density. RI correlates somewhat with physical density (SG) because physical density strongly influences the number of electrons in a media for a specific volume, but it is a separate property. Never use the word 'density' without a qualifier to refer to optical density. Optical density, and therefore refraction, is dependent on wavelength: the shorter the wavelength, the more it is bent by refraction. Violet light is bent more by refraction than red light.

There are several methods of measuring RI, all of which have some relative advantages and disadvantages. Of prime importance to the gemmologist is the method that makes use of a specialized optical instrument known as a *refractometer*, or, as it is sometimes called, the *critical angle refractometer*. This instrument, calibrated to measure directly in refractive indices, is based on the principles of critical angle and total internal reflection. Alternative methods of measurement are discussed at the end of the lesson.

Critical angle and total internal reflection

Reflection and refraction deal with light interacting with a surface, often from the outside. Critical angle deals with light leaving the stone. When

REFRACTIVE INDEX AND MEASUREMENT OF REFRACTIVE INDEX

light travels from a gemstone into air it is travelling from a denser to a rarer medium and the incident ray makes a smaller angle with the normal than the refracted ray. As the angle of incidence increases there will come a point where the refracted ray is at an angle of 90° to the normal. That is, the refracted ray just grazes along the surface between the two media. Any further increase in the angle of incidence will cause the light not to refract but to turn back into the stone according to the laws of reflection. This phenomenon is called *total internal reflection*. *Critical angle* is defined as an angle of incidence on the inside, or optically denser side, of a surface that causes an angle of refraction to form at 90° to the normal. Critical angle marks the beginning of total internal reflection. It separates reflection from refraction inside the stone.

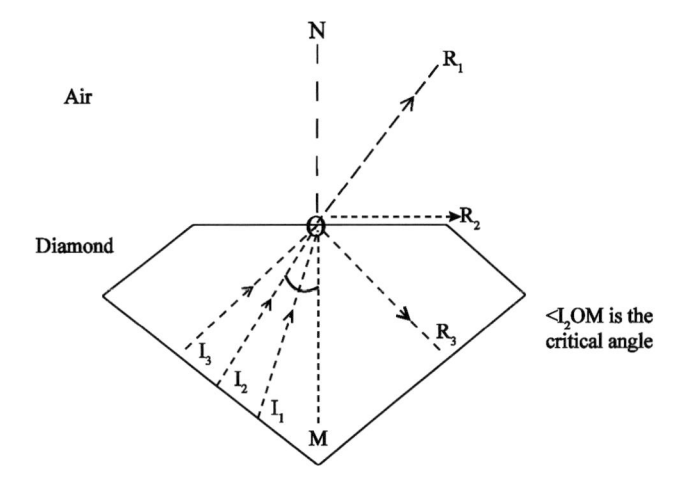

Figure 7.1 Critical angle. Light passing from a gemstone into air is passing from a denser to a rarer medium and is refracted away from the normal (NOM). I_1 is a ray of light incident on the surface and refracted out of the gemstone away from the normal along R_1. As the angle of incidence increases, the angle of refraction increases until an angle of incidence is reached where the refracted ray just grazes the surface ($I_2 \rightarrow R_2$). The critical angle is $<I_2OM$. Any further increase in the angle of incidence will result in light being reflected back into the gemstone as total internal reflection occurs. At I_3, the light will not be refracted out of the gemstone, but will follow the laws of reflection.

$$\text{Sine of the critical angle is equal to } \frac{1}{RI}$$

The size of the critical angle is inversely proportional to refractive index: the greater the RI, the smaller the critical angle. More light will be totally internally reflected in a medium of high refractive index such as diamond, and this explains the brilliance of diamond. Most of the light that enters a well-proportioned round brilliant diamond from the crown is totally internally reflected within the stone, bounced around on the back facets, and returned to exit out through the crown. The smaller the critical angle, the brighter the stone. The RI of diamond is 2.42 corresponding to a critical angle of 24°26'. The RI of quartz is 1.54 and the critical angle is approximately 40°, producing a considerably less bright gem.

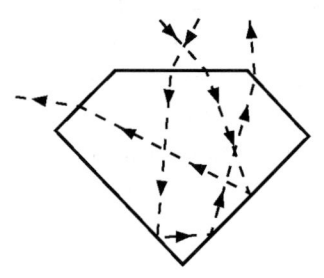

Figure 7.2 Path of light as totally internally reflected in a diamond.

The refractometer is based on total internal reflection, but in this case the refractometer glass is the optically denser medium and we are talking about *total internal reflection inside the refractometer,* not inside the gemstone.

The refractometer

The refractometer is the most valuable analytical instrument available to gemmologists outside of a laboratory. It gives: (a) refractive index; (b) birefringence; (c) optic character; and (d) optic sign. Each will be presented in turn as part of the discussion on taking measurements. We begin with the basics of the instrument.

REFRACTIVE INDEX AND MEASUREMENT OF REFRACTIVE INDEX

Dr G. F. Herbert Smith designed the first practical refractometer, which was marketed in 1907. The modern refractometer, for example the Rayner refractometer, is based on the same principles as the Smith model, but with a system of truncated lenses and prisms that bring the critical angle shadow edge and the scale into a convenient viewing position. The essential feature of the instrument is a hemicylinder or prism of very dense highly refractive glass. Figure 7.3 represents a section through such a glass upon which is resting the test stone, *which must have a refractive index less than that of the glass*. The A incident ray is refracted out through the stone along OA^1 because it falls inside the critical angle. The B incident ray is refracted to a greater degree and passes along the surface as OB^1. This is the incident ray that causes a ray of refraction OB^1 to occur at 90° to the normal. BON is the critical angle. The C incident ray is totally internally reflected at the surface back into the instrument along OC^1. All light incident at greater than the critical angle is totally internally reflected back into the instrument, as represented by the light area on the hemicylinder and the scale. The light visible on the scale is totally reflected at the surface of the liquid connecting the glass to the gemstone; it does not pass into the gemstone first. All light incident at less than the critical angle is refracted out from the instrument into the stone and is lost, as represented by the dark area on the hemicylinder and the scale in Figure 7.3. The demarcation between

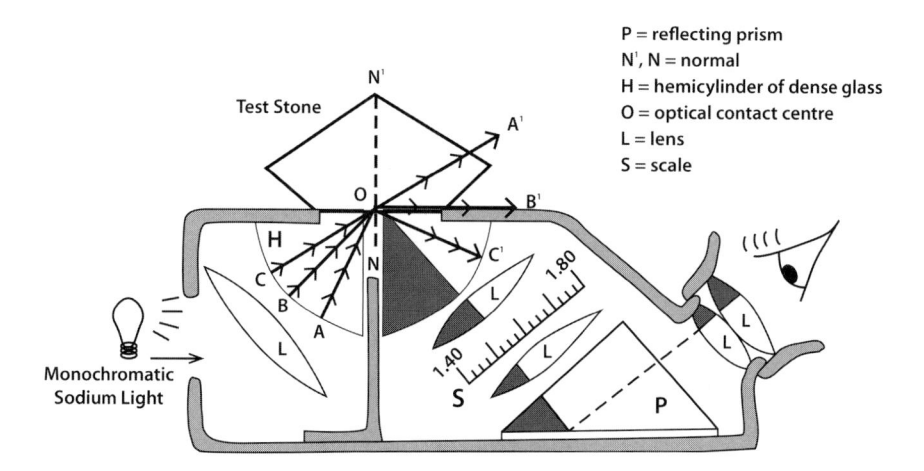

Figure 7.3 Representation of the optic train through a modern refractometer. Where: P is a reflecting prism, N^1N is the normal at the point of incidence, H is the hemicylinder of dense glass, O is the optical contact centre, L is a lens, S is the scale.

light and dark is the shadow edge seen at the critical angle for the gemstone in contact with the refractometer glass. Looking at the scale, one sees a bright area of rays that have been totally internally reflected and a dark area where the rays have been refracted into the stone and do not reach the scale. It should be obvious that if the dense glass is standard, any stone having an RI of 1.72 will have a shadow edge at a different position than a stone with an RI of 1.45. So then, if a lens system, eyepiece and scale are arranged so that the shadow edge can be read, we have a refractometer.

The instrument is limited in its ability to read high refractive indices. To obtain total internal reflection it is necessary for the light to be travelling from an optically denser medium to an optically rarer medium; hence, any stone that has an RI higher than the refractometer glass will not return the rays back into the glass by total internal reflection, but will allow the light to go out through the stone and be lost. The scale will be uniformly dark. The limit imposed by the RI of the glass is typically a maximum of 1.90.

The second limitation on the maximum refractive index reading is that imposed by the contact fluid. When a stone is placed on the glass of the refractometer a film of air remains between the stone and the glass that prevents *optical contact*. To displace this film a liquid, which will *wet* the glass and the stone, has to be placed between them. Of necessity this liquid has to have a higher refractive index than the stone under test for the same reason that the glass must, and thus it forms a second limiting factor. The contact fluid made standard for refractometers has an RI of 1.79, setting the effective upper limit of the refractometer at 1.79. It is a saturated solution of sulphur in di-iodomethane. In the past, a liquid of 1.81 was available. It was a saturated solution of sulphur in di-iodomethane and tetraiodoethylene. This liquid is a suspected carcinogen and is no longer available. There are some higher RI fluids available for periodic use, but all are highly toxic and should be restricted to use in a laboratory. Over the years variations on the instrument have been made, such as those with the dense glass replaced by diamond. Such instruments were capable of measuring a much higher range, but are extremely expensive speciality instruments requiring the use of one of the toxic contact fluids and so are for laboratory use only.

Since RI is a comparison of the speed of light in the gemstone to the speed of light in air, and because optical density, and therefore refraction, is dependent on wavelength, for RI measurements to be meaningful the

REFRACTIVE INDEX AND MEASUREMENT OF REFRACTIVE INDEX

light source must be standardized. By international agreement, the standard light source for measuring refractive index is yellow sodium light with a wavelength of 589.3nm. Sodium light is *monochromatic*, meaning a pure colour with all light energy comprising of one wavelength. Yellow sodium light sources are available with refractometers. A yellow filter over the eyepiece also works but it tends to dim the shadow edges, making them harder to see. If white light is used, the shadow edge appears as a narrow-banded spectrum due to the relative differences in the dispersion of the glass and the dispersion of the gemstone; refractometer glasses are highly dispersive. It is best to use a sodium light as it produces a sharp shadow edge without the distracting rainbow colours. Refractometers are calibrated to measure RI values for sodium light, so if white light is used the reading must be taken where the green passes into the yellow.

Refractometers are expensive and require proper care. The hemicylinder of dense glass is very soft, far softer than the gemstones placed on it, so it is essential to ensure that reckless placing of test stones does not scratch its polished surface. Never take a reading with a dirty stone and never put away the instrument without cleaning off the contact liquid as it can damage the glass if left over longer periods of time. All stones and the glass should be cleaned with soft tissue and alcohol before and after each use. It is also helpful to smear Vaseline on the glass to prevent corrosion if the instrument will not be used for some time.

Testing procedure

The instrument should be placed level on a sturdy surface. A darkened room usually makes for better viewing of the shadow edges. Be certain that the stone and the refractometer glass are thoroughly clean before testing. Place a tiny drop of contact liquid in the centre of the refractometer glass. Place the stone table facet down on top of the liquid on the glass and close the cover. View the scale through the eyepiece. If the instrument has an internal scale that is not in focus, adjusting the eyepiece can focus it. If the entire field of view is dark, the stone has an RI higher than the limit of the refractometer. This is termed *negative RI*. It happens with: diamond and its imitators: cubic zirconia, GGG, YAG, strontium titanate, synthetic rutile, synthetic moissanite; as well as demantoid garnet, sphene and some others.

If the stone is within refractometer range, the upper part of the scale with lower numbers will be seen dark and the lower part with higher numbers will be bright. Observe the shadow edge(s) at the light/dark boundary and turn the stone carefully to observe any movement. It is important to understand that the direction of testing is *along the surface of the facet* placed on the glass, *not into the facet*. Therefore it is possible to test various directions by rotating the stone. When turning the stone be sure not to move it along the glass as this results in inaccurate readings. Take recordings of the readings, pairs of readings if there are two shadow edges, as the stone is turned eight or more times in one 180° rotation. At exactly 180° you will be looking along the same line in the opposite direction, so the readings should be the same as at 0°. Take all readings to the third decimal place, which must be estimated, to determine birefringence. Gemstones from the isometric system have no birefringence, as a single shadow edge is all that is seen. In uniaxial stones and biaxial stones, birefringence is the maximum difference between the higher shadow edge and the lower shadow edge.

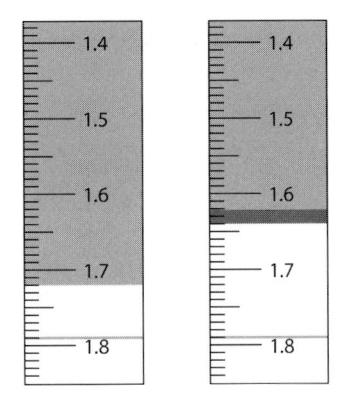

Figure 7.4 Refractometer scales showing sample readings. Left: a single shadow edge at 1.72. This is the isotropic gemstone spinel. Right: two shadow edges seen at 1.622 and 1.640, giving a birefringence of 0.018. This is the anisotropic gemstone tourmaline. The shadow edge at 1.79 is that of the contact fluid.

The Rayner Dialdex model has no scale and the shadow edges are seen on a plain field. Measurements are taken by turning a drum dial at the side of the instrument. As this dial, marked in refractive indices, is turned, a black 'ribbon' is seen to travel down the field of view. When the edge of this 'ribbon' is aligned with the shadow edge, the reading is taken off the dial. Some find this instrument better to work with as the external scale allows for easier readings and a better estimation of the third decimal place.

For good refractometer readings one must: (a) have flat polished surfaces on the stone and refractometer as damaged refractometer glass will dim readings; (b) clean the stone and the refractometer glass; (c) use RI fluid with no exolved crystals; and (d) only use a small drop of contact fluid and make sure the stone is centred on the glass.

The shadow edge(s) seen from testing any one stone depend on: refractive index, birefringence, optic character, optic sign, and the slope of the test facet in relationship to the structure of the stone. To understand readings, one must understand the concepts of isotropy, anisotropy and birefringence; optic axis; and uniaxial and biaxial.

Isotropy and birefringence

Isotropy and anisotropy are often referred to as single refraction and double refraction respectively. But since anisotropic refers to all optical properties, including colour, absorption spectra, etc., it is best to learn to speak of these concepts by their proper terms: isotropic and birefringent. Singly refracting and doubly refracting are misnomers. A gemstone that is elsewhere referred to as *singly refracting* is properly termed *isotropic* and a gemstone that is elsewhere referred to as *doubly refracting* is properly termed *birefringent*. Light striking a gemstone is ordinary unpolarized light containing rays vibrating in random directions at right angles to the direction of travel. To envision this, think of spokes emanating from the centre of a bicycle wheel. When light enters an isotropic gemstone the light remains as one ray and is refracted as one ray, producing only one refractive index. This refractive index will be invariable when the stone is turned. If there is any doubt that one is seeing a single shadow edge, a polarizing filter can help. Place the filter over the eyepiece and rotate it. If the questionable edge is in fact two separate edges they will alternatively blink on and off but they do not move. If the stone is genuinely isotropic no difference will be seen. Isotropic materials are minerals belonging to the cubic or isometric system, and amorphous substances such as glasses, resins and all liquids.

In isotropic materials no matter which way the light travels, the vibration directions are all presented with the same structure and this structure does not change the light in any way with the exception of velocity: it slows or speeds up on entering an optically denser or rarer medium respectively.

When light enters an anisotropic gem, the ordinary unpolarized light is resolved into two separate rays, each polarized to vibrate at 90° to the other; meaning these rays vibrate in one plane only and the plane of vibration of one ray is 90° to the plane of vibration of the other ray. These two plane polarized rays are travelling at different velocities, producing two refractive indices – one for each ray. Usually they travel in different directions because, if they refract, each is refracted to a different degree. For example, the refractive indices of quartz are 1.544 and 1.553.

Birefringence

Double refraction divergence, the angular separation of the two refracted plane polarized rays with two separate RIs, is a property of all crystal systems except the isometric system. Birefringence is the maximum difference between the two refractive indices in an anisotropic gemstone. With any anisotropic stone, no matter how the facet is sloped, there will always be one direction where the lowest possible RI can be seen and one direction where the highest possible RI will be seen, so full birefringence can be measured. For all stones, uniaxial or biaxial, the difference between maximum and minimum readings obtained on any one facet must always represent the true birefringence of that stone.

Using a lens it may be possible to see the effects of birefringence, the double refraction divergence, in stones with a large birefringence such as zircon (0.059) and peridot (0.036). This effect is seen as a doubling of the back facet edges when viewed through the stone. This is provided that one is not looking along an optic axis where the double refraction divergence is not seen, as birefringence is zero in this direction. This observation is useful for differentiation in some instances such as with diamond, where there is no double refraction divergence because it is isotropic, and zircon where the double refraction divergence is easily visible in other than the direction of an optic axis. Experience also shows that doubling of the back facets may be seen in stones of low birefringence under higher magnification. There are several gemstones in the RI range of 1.62–1.63 but their birefringence varies: tourmaline 0.018, topaz 0.10 and apatite 0.002. Under higher magnification it may be possible to estimate the degree of doubling as an aid to identification.

Optic axes

In anisotropic gemstones, there are one or two directions parallel to which it is not possible to observe birefringence. These directions are termed *optic axes*. Optic axes are often referred to as directions of single refraction in a doubly refracting mineral, but this is incorrect in the strict sense of true optic science. It is more correct to say that an optic axis is a direction through a birefringent stone where light travelling in this direction shows zero birefringence: uniaxial denotes one such axis;

Table 7.1 Refractometer key points

The refractometer allows one to observe a shadow edge at the critical angle when refraction into the stone ceases and total internal reflection into the refractometer begins.

All light passing inside the critical angle is refracted out through the stone.

All light passing outside the critical angle is totally internally reflected back into the instrument.

There are two things that can limit shadow edge movements:
 (1) The structure of the stone.
 (2) The slope of the facet in relationship to the structure of the stone.

For all stones, uniaxial or biaxial, the difference between maximum and minimum readings obtained on any one facet must always represent the true birefringence of that stone.

If two shadow edges are seen and neither move, the stone *must* be uniaxial but it is not possible to determine optic sign without using a polarizing filter.

If two shadow edges are seen and *both* move, it is not possible for the stone to be uniaxial; it *must* be biaxial.

biaxial denotes two such axes. The tetragonal, hexagonal and trigonal systems are uniaxial with the optic axis parallel to the principal crystal axis. The orthorhombic, monoclinic and triclinic systems are biaxial with the optic axes in no particular relation to the crystallographic axes. It is important to understand that crystallographic axes have a particular location – for example, the principal crystal axis is in the centre – but optic axes are *directions,* not particular locations. An optic axis is in all parallel directions, not just in the centre.

Uniaxial

In uniaxial stones, one of the rays has a refractive index that is constant and so its shadow edge never moves. Along the optic axis in the tetragonal, hexagonal and trigonal systems the vibration directions are all presented with the same structure and this structure does not change the light in any way with the exception of velocity. It obeys the ordinary laws of refraction and is termed the *ordinary ray*. It is usually denoted by the letter o, and the RI value is named as Greek letter omega = ω. The other ray, termed the *extraordinary ray*, has a refractive index that varies according to its direction of vibration, which is limited by the direction of propagation. Its value varies from that of the ordinary ray to a second limiting value. The extraordinary ray is denoted by the letter e, and the RI value at the second limit is named as Greek letter epsilon = ε. At right angles to the optic axis, the RI of the extraordinary ray is at its maximum difference from the ordinary ray and the value of epsilon (ε) is taken at this limit. At angles in between, the variable RI will take an intermediate value but the value of epsilon (ε) is still that of the extreme reading away from omega (ω). If the extraordinary ray has an index of refraction greater than that of the ordinary ray, the optic sign is said to be positive; and if the extraordinary ray has an index of refraction less than that of the ordinary ray, the optic sign is negative.

If two shadow edges are seen and *neither moves,* the stone *must* be uniaxial but it is not possible to determine optic sign without using a polarizing filter because it is not known which is the extraordinary ray. This occurs with the optic axis vertical and is typical of tourmaline because the darkest colour in tourmaline is along the principal crystallographic axis so the table is typically cut perpendicular to the optic axis, therefore neither shadow edge moves. Polaroid sunglasses

REFRACTIVE INDEX AND MEASUREMENT OF REFRACTIVE INDEX

UNIAXIAL

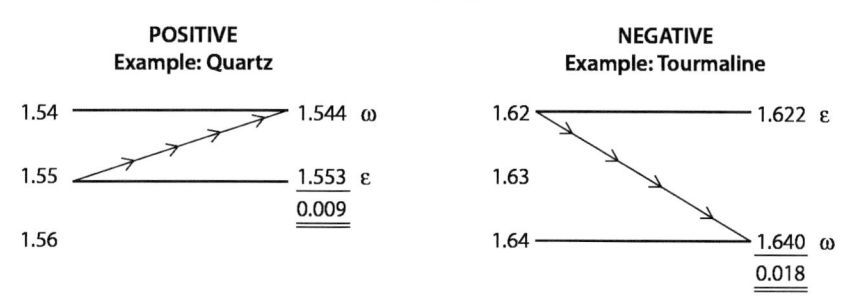

POSITIVE
Example: Quartz

1.54 ─────────── 1.544 ω
1.55 ─────────── 1.553 ε
1.56   0.009

NEGATIVE
Example: Tourmaline

1.62 ─────────── 1.622 ε
1.63
1.64 ─────────── 1.640 ω
                   0.018

BIAXIAL

POSITIVE
Example: Diopside

1.67 ─────────── 1.670 α
1.68 ─ ─ ─ ─ ─ ─ 1.680 β
1.69
1.70 ─────────── 1.700 γ
                   0.030

NEGATIVE
Example: Sinhalite

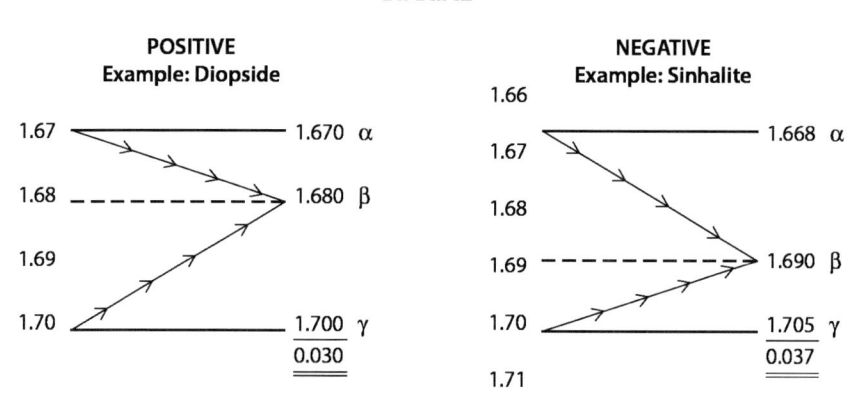

1.66
1.67 ─────────── 1.668 α
1.68
1.69 ─ ─ ─ ─ ─ ─ 1.690 β
1.70 ─────────── 1.705 γ
1.71   0.037

Figure 7.5 Illustrative refractometer readings for uniaxial and biaxial stones. Top left: quartz is uniaxial positive with a birefringence of 0.009. The 1.544 reading is the invariable ordinary ray and the 1.553 reading is the extraordinary ray that varies from this maximum reading downwards to 1.544. Top right: tourmaline is uniaxial negative with a birefringence of 0.018. The 1.640 reading is the invariable ordinary ray and the 1.622 reading is the extraordinary ray that varies from this minimum reading upwards to 1.640. Bottom left: diopside is biaxial positive with a birefringence of 0.030. The 1.670 reading is alpha that varies from this minimum reading upwards to beta, the 1.680 reading is beta, and the 1.70 reading is gamma that varies from this maximum reading downwards to beta. Bottom right: sinhalite is biaxial negative with a birefringence of 0.037. The 1.668 reading is alpha that varies from this minimum reading upwards to beta, the 1.690 reading is beta and the 1.705 reading is gamma that varies from this maximum reading downwards to beta. A simple rule to determine optic sign is: if the higher RI has greater variability the optic sign is positive, and if the lower RI has greater variability the optic sign is negative. These are quick diagrams to remember for exam questions.

block horizontal vibrations and transmit vertical vibrations, so viewing the shadow edges through the glasses will show only the extraordinary ray and then the optic sign can be determined. If the vibration direction is vertical, it is the one that *would* move if it were not locked in place by the slope of the facet.

Uniaxial possibilities

There are five possible scenarios. Quartz is used as the uniaxial positive example and tourmaline as the uniaxial negative example.

When the test facet is cut parallel to the optic axis one of two things can happen:

1) The lower RI edge is stationary at 1.544 and is visible in all directions. The higher RI edge varies from its maximum position all the way down to the lower RI edge (1.553→1.544). The birefringence is 0.009 and the optic sign is positive.
2) The higher RI edge is stationary at 1.640 and is visible in all directions. The lower RI edge varies from its minimum position all the way up to the higher RI edge (1.622→1.640). The birefringence is 0.018 and the optic sign is negative.

When the test facet is cut perpendicular to the optic axis there is one outcome:

3) The lower RI edge is stationary at 1.622 and is visible in all directions. The higher RI edge is stationary at 1.640 and is visible in all directions. The birefringence is 0.018, but since one edge is limited completely by slope it is not possible to determine the optic sign until a polarizer is used to block the horizontal vibrations. It will show which edge would be moving if not for the slope limitation; it shows the extraordinary ray. In this case it would show 1.622. The optic sign is negative and the stone is tourmaline. In the case of quartz, it would show 1.553 because the optic sign is positive. No matter how the stone is rotated, one vibration direction will be perpendicular to the optic axis and one vibration direction will be parallel to the optic axis and the two shadow edges remain locked.

REFRACTIVE INDEX AND MEASUREMENT OF REFRACTIVE INDEX

When the test facet is cut at any degree of inclination to the optic axis there are two outcomes:

4) The lower RI edge is stationary at 1.544 and is visible in all directions. The higher RI edge varies from its maximum position partway downward to some limit (1.553→1.549). The stone is uniaxial positive with a slope limitation.
5) The higher RI edge is stationary at 1.640 and is visible in all directions. The lower RI edge varies from its minimum position partway up to some limit (1.622→1.631). The stone is uniaxial positive with a slope limitation. In this instance the test facet is at 45° to the optic axis as the extraordinary ray moves exactly halfway to the ordinary ray.

With two moving edges it is not possible for the stone to be uniaxial; it *must* be biaxial.

Biaxial

In biaxial stones, both shadow edges move. There is no ordinary ray; both rays have refractive indices that vary according to vibration direction. It should be made absolutely clear that it is vibration direction that influences RI, not direction of propagation. Two numbers can describe each of the two rays: the numerically upper and lower limits of variability in each case. It is not necessary to have four values to describe biaxial stones because the highest limit of one variable RI always coincides with the lowest limit of the other variable RI, and so duplicates its value. The lowest limit of the lower variable RI is designated as alpha, denoted by the Greek letter α. The shared limit of variability is designated as beta, denoted by the Greek letter β. And the highest limit of the higher variable RI is designated as gamma, denoted by the Greek letter γ.

The RI of the lower shadow edge will vary between alpha, its extreme lowest reading, upwards to the intermediate value of beta. The RI of the higher shadow edge will vary between gamma, its extreme highest reading, downwards to the intermediate value of beta. Beta represents the RI along the optic axis, the direction in which there is no birefringence, and the observation is that of one shadow edge. Neither shadow edge can pass beyond the limit of beta. The value of beta is the single reading which sets the higher limit of the shadow edge moving from alpha and the lower limit of the shadow edge moving from gamma. The optic sign of biaxial stones is determined by whichever edge has the greater variability: if gamma is farther from beta than alpha is, the optic sign is positive; if alpha is farther from beta than gamma is, the optic sign is negative. It is not possible to tell optic sign from just alpha and gamma readings. Some information about beta is needed.

One will always be able to see alpha and gamma regardless of the slope of the facet but there may be some confusion with beta. Depending on the slope of the test facet to the vibration directions of the stone it is possible that the shadow edge moving from alpha or gamma does not

ISOTROPIC   **UNIAXIAL**   **BIAXIAL**

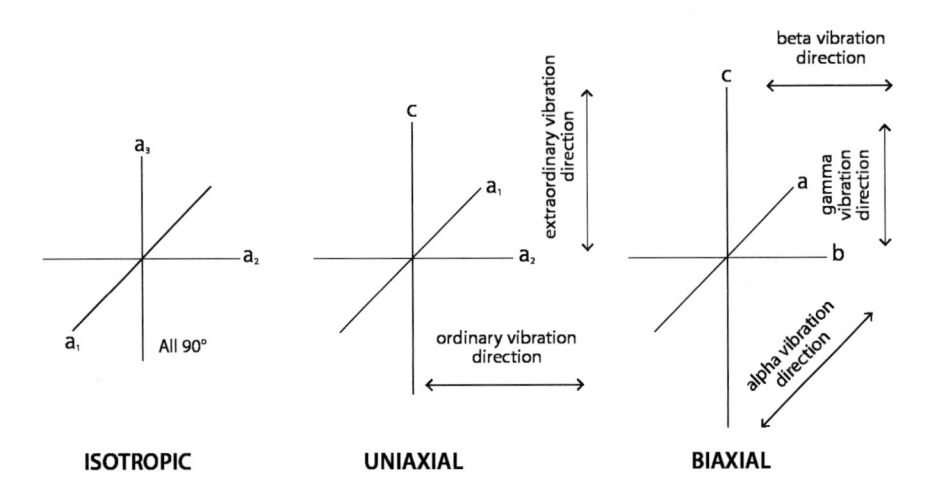

Figure 7.6 Axes and vibration directions. Left: in the isotropic system all axes are equal and all directions are the same. Light is not changed in any way when it is transmitted through isotropic materials except for velocity. Middle: in uniaxial systems, **c** is usually unequal to **a** so there are two different kinds of directions. Light travelling down **c** encounters a_1 and a_2, which are the same, so in this direction light is not changed in any way except for velocity and no birefringence or double refraction divergence is seen. This is the direction of the optic axis. In directions other than this there are two plane polarized rays with two different RI values. Right: in biaxial systems, **c** is usually the longest and **a** is usually the shortest, with **b** somewhere in between, so there are three different kinds of directions and birefringence or double refraction divergence is seen in all directions except in the direction of an optic axis where light is not changed in any way except for velocity.

REFRACTIVE INDEX AND MEASUREMENT OF REFRACTIVE INDEX

reach all the way to beta. This means there will be two intermediate readings between the maximum and minimum readings, not one single reading. It is important to understand, as this is often expressed incorrectly, that beta is not somewhere in between the two intermediate readings. If the slope of the facet is blocking one edge from reaching beta it cannot be blocking the other. One of the edges must move all the way to beta. The beta value is not an average of the two readings: one is *true* beta and one is *false* beta. Polaroid sunglasses can differentiate between the two readings because they transmit vertical vibration directions. Check each intermediate limit with Polaroid sunglasses or with a polarizing filter oriented to block reflected glare. False beta will be visible when viewed through the sunglasses and true beta will disappear. If it is not visible, then it is true beta. If one edge travels more than halfway between alpha and gamma, then this edge determines optic sign without knowing the exact value for beta, as it is not possible for the other edge to have greater variability.

Biaxial possibilities

There are four possibilities. Peridot is used as the example with alpha at 1.650, beta at 1.667, and gamma at 1.686. The optic sign is positive because gamma is farther from beta: gamma to beta is 0.019 and alpha to beta is 0.017. With a stone of negative optic sign, such as sinhalite, it may have alpha at 1.670, beta at 1.695, and gamma at 1.708. In this case, alpha to beta is 0.025 and gamma to beta is 0.013, and alpha is farther from beta, but the following still holds:

1) When the test facet is cut perpendicular to the vibration direction for alpha: the lower RI edge is stationary at alpha 1.650 and is visible in all directions. The higher RI edge at its maximum is gamma 1.686 and it moves downward to beta 1.667.
2) When the test facet is cut perpendicular to the vibration direction for gamma: the higher RI edge is stationary at gamma 1.686 and is visible in all directions. The lower RI edge at its minimum is alpha 1.650 and it moves upwards to beta 1.667.
3) When the test facet is cut perpendicular to the vibration direction for beta: the intermediate reading of beta 1.667 is visible in all directions. The lower RI edge moves upwards from alpha 1.650 to beta 1.667 at

which point only one shadow edge is seen. As the rotation continues, the higher RI edge moves from beta 1.667 upwards to gamma 1.686.
4) When the test facet is inclined to some degree: both edges move. As the stone is rotated there will be one position where the lower RI edge is at its minimum, alpha 1.650, and there will be one position where the higher RI edge will be at its maximum, gamma 1.686. The lower RI and higher RI edges vary upwards and downwards respectively towards beta 1.667, but may not actually simultaneously be at their extreme location from beta and/or may not simultaneously coincide with beta. Thus, there may be two intermediate readings: the two possible betas, such as 1.667 and 1.675. One of the intermediate readings must be true beta because a slope limit can restrict one of the edges from reaching beta but not both. Polaroid sunglasses transmit vertical vibration and show *false* beta. The edge that disappears is true beta.

Distant vision readings

The refractive index of stones with curved surfaces may be obtained with some accuracy by using the distant vision, or spot reading, method. In this method one is looking at the whole optical contact area instead of a shadow edge. It is mandatory to limit the size of the drop of contact fluid as the size of the contact spot dictates accuracy. The contact from a curved surface on a flat surface is really only a point, so if a big drop of contact fluid is used it will climb up on the sides of the stone and these areas will give erroneous readings. A good method is to apply an ordinary small drop, touch the stone to the drop, and then lift the stone and wipe away what is left of the drop, using only the tiny amount adhered to the stone. Place the stone in the centre of the refractometer glass. With the eye held about 30–40 cm away from the eyepiece, move the head slightly until a position is found where it is possible to see the point of contact of the liquid and the stone as a small disc in the middle part of the scale. If one is working with a refractometer with an internal scale and if the contact spot is viewed against index readings that are below the RI of the stone, it will appear completely dark. As the line of vision moves towards the higher index readings, at a certain point the contact spot will move from dark to light as total internal reflection sets in, providing the stone has an RI within refractometer range. At some intermediate position between the two, a line of shadow will bisect the disc. It will be half light and half dark. When this position is found, lower the eye towards the eyepiece so the scale can be read. If needed, looking through an opaque material (such as cardboard) with a pinhole punched in it will help bring the scale into focus. If the refractometer is a Dialdex it is easiest to bring the ribbon down first and then slowly raise it up until the light/dark bisects the spot and at this point the reading can be taken off the dial. Accuracy to three decimals cannot be achieved with spot readings. It is not necessary to use sodium light. The method requires practice, the two edges of anisotropic stones are rarely seen, and readings are often slightly low. The method is also useful for small stones with tiny table facets.

Alternate methods of determining RI

The refractometer is the gemmologist's most important means of measuring RI because of its accuracy, but it is not the only method. RI can be determined by direct measurement and Brewster angle meter, as well as estimated by various immersion methods.

Direct measurement

Direct measurement, also known as the real and apparent depth method, is useful for transparent polished stones. It requires a microscope with a calibrated focusing scale or the use of measuring callipers. RI can be approximated by dividing the stone's real depth by its apparent depth when measured with the focus through the stone.

Position the stone with the culet in contact with the microscope stage and the table parallel to the stage. Plasticine is a simple means of securing the stone. Under high magnification, focus on the table and read the position of the focus from the scale (T). Next, lower the focus through the stone until the culet is in sharp focus and take another reading of the focus from the scale (C). Subtract the culet focus reading from the table focus reading to determine the apparent depth of the stone (T − C = apparent depth). Most microscopes do not have a calibrated focusing scale. In this case, measuring callipers can be used to determine how much the head of the microscope moves when the focus is adjusted from the table to the culet. This measurement is the apparent depth of the stone. Finally, determine the real depth of the stone by either measuring it with the callipers or by removing the stone and focusing the microscope on the stage and taking a reading of the focus (S). Subtracting this reading from the table focus reading gives the real depth (T–S = real depth). RI is calculated by simply dividing:

$$\frac{\text{real depth (T - S)}}{\text{apparent depth (T - C)}} = RI$$

The method is limited to accuracy of +/- 1%, but can be used for high RI stones that are over the limit of the refractometer.

Brewster Angle meter

Scottish physicist Sir David Brewster formulated the rule governing polarization by reflection. Brewster's law states that when monochromatic light is reflected from the flat surface of an optically denser medium, that reflected light is completely horizontally polarized in the plane of that surface when the reflected and refracted rays are perpendicular (at 90°) to one another. In other words, at the Brewster Angle, only horizontally polarized light is reflected from the surface of the optically denser medium. Measurement of the Brewster Angle provides a convenient means by which to measure the RI of a gemstone: RI is simply the tangent of its Brewster Angle. Any standard calculator with trigonometric function can be used for the conversion. Simply enter the Brewster Angle for 589.3nm yellow sodium light, say 67.53, press the 'tangent' button and read the related RI – 2.417 (2.42) – directly from the calculator. In this case the test stone is diamond.

B.W. Anderson established the diagnostic value of the Brewster Angle in 1941, but practical difficulties prevented him from developing a usable instrument. Peter G. Read developed a portable, commercial battery-operated meter available through GAGTL. It uses a 5 milliwatt 670nm vertically polarized laser light source. The stone is placed table-down over the test aperture and a control knob rotates the laser beam around the test aperture. The intensity of the reflected ray is viewed on

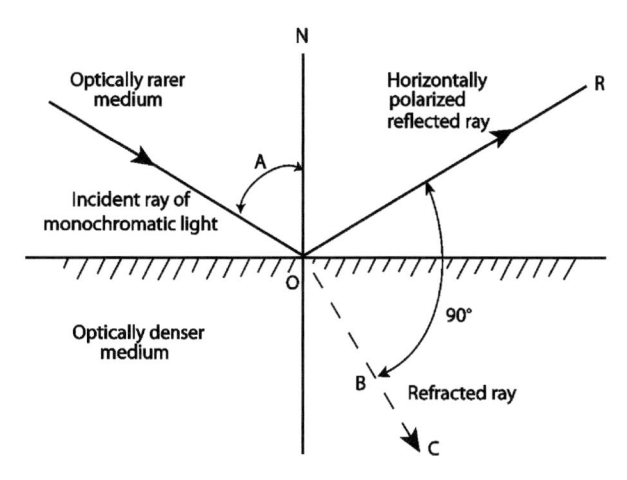

Figure 7.7 At the Brewster Angle, only horizontally polarized light is reflected from the surface of the optically denser medium. Angle A = Brewster Angle.

a translucent screen. Since the laser beam is vertically polarized, and at the Brewster Angle only horizontally polarized light is reflected from the surface, when the laser beam reaches the Brewster Angle the intensity is reduced to a null, which is displayed as a dark horizontal bar. At this point the Brewster Angle can be read directly from the scale on the control knob. The Brewster Angle meter provides diagnostic readings for gemstones over a far greater range than standard refractometers. It covers angles from 54° to 72°, representing the refractive index range 1.38 to 3.08; it is not overly sensitive to the surface condition of the stone; and requires no contact fluid. The instrument is calibrated to measure the Brewster Angle at 670 nm, rather than at the standard 589.3 nm, and a gemstone table is provided with corresponding Brewster Angles for the 670 nm wavelength. It cannot measure birefringence and can only give the RI for the horizontal vibration, which will be somewhere within the range of RI readings for the stone being tested.

Immersion methods

Immersion methods use high refractive index liquids and should never be performed on porous stones such as opal, or stones that may be damaged or softened by the liquids, such as amber. With any stone it is best to limit exposure to the liquids by testing quickly and cleaning the stone.

Table 7.2 Fluids for immersion methods

Name	RI
Water	1.334
Alcohol	1.36
Olive Oil	1.47
Bromoform	1.59
Monobromonaphthalene	1.66
Di-iodomethane (methylene iodide)	1.74
Di-iodomethane with sulphur	1.79

Becke method

This method is best suited to thin fragments or chips. Put the chip into an immersion cell and cover it with a liquid of known RI. Place the cell on the stage of a microscope and in transmitted light focus on the chip. When the microscope tube is raised from the position of exact focus the bright line that was seen at the margin of the specimen and the liquid will travel into the medium of higher RI. When the focus is lowered, the bright line moves into the medium of lower RI.

Modified Becke method

The Becke method has been adapted to faceted gemstones by R. K. Mitchell. Immerse the stone, table facet down, in an immersion cell of liquid of known RI and place it on the microscope stage. Adjust the microscope lighting so that transmitted light is restricted to the area of the gemstone. If the facet edges change from *light* to *dark* as the focus is lowered from the liquid into the stone, then the RI of the stone is higher than that of the liquid. If the facet edges change from *dark* to *light* as the focus is lowered

Table 7.3 Refractive index

Listed are the average or mean values for the refractive index of various gemstones. Many gemstones show greater or considerable variation by colour and variety, and as a result of impurities. Values separated with '–' denote birefringence, those with 'to' indicate a single value that falls within that range. The right-hand column shows optic character: isotropic, uniaxial or biaxial; and optic sign.

Gemstone	RI	BI	Character/Sign
Fluorite	1.43	–	Isotropic
Opal	1.45	–	Isotropic
Amber	1.54	–	Isotropic
Quartz	1.544–1.553	0.009	Uniaxial+
Beryl	1.57–1.58	0.006	Uniaxial–
Topaz – white and blue	1.610–1.620	0.010	Biaxial+
Nephrite	1.62	0.027*	Biaxial–
Tourmaline	1.62–1.64	0.018	Uniaxial–
Topaz – brown and pink	1.63–1.64	0.008	Biaxial+
Andalusite	1.63–1.64	0.010	Biaxial–
Peridot	1.65–1.69	0.036	Biaxial+
Jadeite	1.66	0.013*	Biaxial+
Spodumene	1.66–1.68	0.015	Biaxial+
Sinhalite	1.67–1.71	0.038	Biaxial–
Spinel	1.72	–	Isotropic
Pyrope garnet	1.73 to 1.76	–	Isotropic
Hessonite garnet	1.72 to 1.74	–	Isotropic
Chrysoberyl	1.74–1.75	0.009	Biaxial+
Corundum	1.76–1.77	0.008	Uniaxial–
Almandine garnet	1.77 to 1.81	–	Isotropic
Demantoid garnet	1.89	–	Isotropic
Zircon	1.93–1.99	0.059	Uniaxial+
Strontium titanate	2.41	–	Isotropic
Diamond	2.42	–	Isotropic
Synthetic rutile	2.62–2.90	0.287	Uniaxial+

* Usually undeterminable

Note: Glass (paste stones) may have any refractive index between 1.50 and 1.70, and are isotropic. Any stone giving a single reading in this range should arouse suspicion.

from the liquid into the stone, then the RI of the stone is lower than that of the liquid. If necessary, repeating the procedure with a series of liquids of known RI can approximate the RI of the stone more precisely.

Contrast immersion

This is useful for most stones, including those that are small or mounted. The method is based on the fact that when a specimen is immersed in a liquid having a similar RI to itself the relief is low, meaning the edges tend to disappear or, if the stone is strongly coloured, its facet outline becomes indistinct. The specimen is immersed in one liquid after another until one is found in which the stone most completely disappears. It is then known that the specimen has an RI approximating to that of the liquid. Such a test may be convenient for testing a cluster ring containing diamonds and suspected synthetic white sapphire or synthetic white spinel. In this case it is not necessary to obtain an accurate approximation of RI; it is only necessary to immerse it in a liquid such as di-iodomethane (RI 1.74). The diamonds would stand out in high relief and either of the synthetic imitations would disappear. This test will not work for differentiation between diamond and its higher RI imitations such as strontium titanate and synthetic moissanite.

LESSON 8

Polarized Light and Colour in Gem Distinction

An ordinary ray of light vibrates in random directions at right angles to the direction of travel. In contrast, polarized light occurs with specific vibration directions. Most instances of polarized light are called plane polarized (more properly termed 'linear polarized'), with one direction of vibration 90° to the direction of travel. The optical properties of plane (or linear) polarized light are the basis for gemmological testing with the polariscope. Colour is a sensory perception that gives us much of the beauty of gemstones and the colour nuances exhibited by various gemstones provide a degree of assistance in gemstone recognition to the expert, but colour alone can never be an absolute means to identification. Reliance on instrument testing is necessary in all cases. The spectroscope, dichroscope and Chelsea filter are instruments for gemstone analysis based on colour. The dichroscope is based on both polarized light and colour.

Producing polarized light

Light can be polarized by several methods. In any direction of travel that is not an optic axis direction, anisotropic gemstones polarize light by splitting the light into two rays plane polarized to vibrate at right angles to each other; and both of these planes are perpendicular to the direction of travel.

The Nicol prism is historically significant as the first use of calcite to produce polarized light, where the ordinary ray is eliminated and a single plane-polarized ray, the extraordinary ray, emerges as polarized light. A

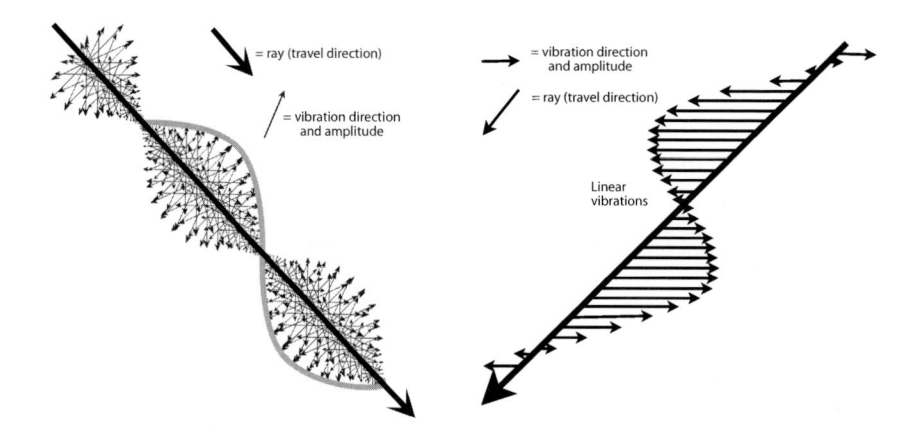

Figure 8.1 On the left: ordinary unpolarized light in which vibration directions are random. On the right: plane-polarized light that vibrates in one plane only at 90° to the direction of travel. In both cases, amplitudes establish the wave shape.

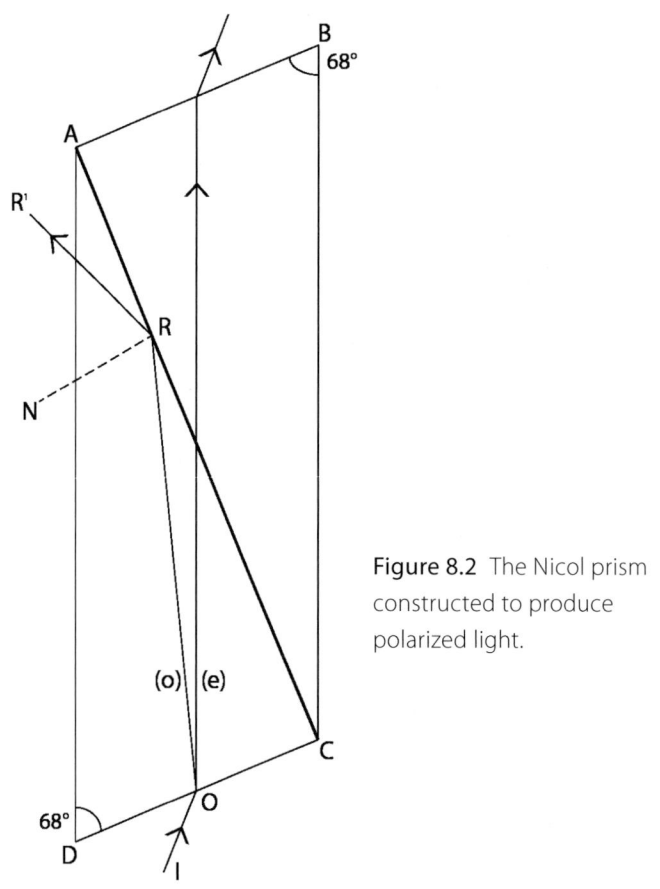

Figure 8.2 The Nicol prism constructed to produce polarized light.

Nicol prism is made from a cleavage rhomb of optically clear calcite, or Iceland Spar as it is sometimes called, which is about three times longer than it is wide. The natural angles ABC and ADC are 70.53° and must be ground and polished down to 68°. The rhomb is cut along AC, the shorter diagonal, and the halves are polished and cemented back together with Canada balsam. Calcite has a very large birefringence, 0.172, and when a ray of ordinary light enters the rhomb the incident ray IO is split into two widely separated plane-polarized rays. Omega, the RI of the ordinary ray, is 1.658, which is significantly higher than the 1.55 RI of the Canada balsam, and it strikes the balsam film at an angle greater than the critical angle, so is totally internally reflected out of the crystal along RR^1 and is absorbed by the non-reflecting mounting of the prism. Epsilon, the RI of the extraordinary ray, is 1.486 and since it is incident on the film at less than the critical angle, it is refracted into the layer of balsam and is able to pass through the rhomb and emerge as a plane-polarized ray.

Polarizing filters, *Polaroid*, is a manufactured transparent plastic embedded with a vast number of sub-microscopic crystals of quinine iodosulphate subjected to magnetic force in order to orient the crystals so that their axes are aligned with one another. Another method is by control of the molecular structure of the plastic. This is produced by forming a brush-like structure inside a sheet of polyvinyl-alcohol. It is stretched out so the molecules untangle and align themselves parallel to the direction of the stretch. It is then dipped in iodine solution to activate the polarizing properties. Polaroid passes light with minimum attenuation when it is vibrating in one plane only and increasingly absorbs rays that are vibrating at increasing angles to this plane.

Light can also be partly polarized by reflection off a smooth flat surface, an effect commonly seen on a lake. This is the basis for orienting a polarizing filter for use with the refractometer where it is set to block the reflected glare, the horizontal vibrations such as that from a desk, and therefore transmit only vertical vibrations. The maximum degree of polarization when light is reflected from the surface of a transparent substance is attained when the reflected and refracted rays are at 90° to each other. This is the Brewster Angle.

Polariscope

The polariscope is a simple instrument constructed of two linearly polarizing media, usually Polaroid film, protected by strain-free glass plates held in a frame overtop of a built-in light source. The bottom polarizing medium, the one closest to the light, is called the polarizer. A separate glass plate above the polarizer rotates such that stones placed on it can be turned 360° for testing. The top polarizing medium, called the analyser, is spaced above the polarizer with sufficient room to insert stones for testing.

The light from the light source vibrates in all directions. The polarizer restricts it to vibrating in only one direction, such as ↕ (N–S). With the analyser in this same orientation, light polarized in this plane (↕) will pass through and be seen through the top of the instrument, but if the analyser is turned 90° to the polarizer, as in ↔ (E–W), all transmission of light is blocked. This is the *extinction position*. It is the working position of the polariscope. At any other orientation to each other, the vibration direction of the light from the polarizer is vectored to pass through the analyser; the degree to which it passes depends on the degree to which the polarizing media are offset from the extinction position with maximum transmission when they are in parallel orientation. Thus, some light is always transmitted through a polariscope except when it is exactly in the crossed, or extinction, position. This principle is the basis for testing.

The polariscope's primary purpose is to determine whether materials are isotropic, anisotropic or polycrystalline. It can only be used with transparent or translucent materials although it does not matter whether they are polished or unpolished. It does not give results for opaque materials and is often unreliable with semi-translucent materials. To test a stone, place it on the glass plate of the polarizer and rotate it 360° while observing it through the analyser. If the stone remains dark throughout the entire rotation it means that the polarized light from the bottom filter was passed through the stone without the stone influencing its wave shape, therefore no light was vectored to pass through and the extinction remained unaltered. This happens with all isotropic materials including amorphous materials such as glass, plastics and liquids. If the stone turns light/dark as it is rotated, then it is anisotropic. When such a stone is dark it is oriented in a position with its vibration directions exactly parallel to the vibration directions in the polars. When the stone's vibration directions are oblique, the light from the polarizer is vectored into these diagonal directions of vibration and can be transmitted through the stone. Upon exiting the stone, the light is vectorized by the analyser and transmits through to be seen by the eye. There are four positions of extinction, where each of the two directions in the stone aligns with each of the two directions in the polars. At 45° to this, there are four positions of maximum brightness, so the stone turns light/dark four consecutive times in one 360° rotation. This observation is made with all anisotropic stones except when looking in the direction of an optic axis. Stones are typically tested table-down, and if this reads as isotropic then turn the stone to test it on a pavilion facet. One must test at least three positions to confirm anisotropy, as there are two optic axes in biaxial stones.

There are some cautions. Garnets can give unreliable results. And with well-cut stones of high RI, remember that they are cut so that all the light that enters the crown is totally internally reflected to exit the crown, so if such a stone is tested table-down, it may appear uniformly dark. For this reason zircon, an anisotropic stone, can appear dark throughout the rotation and may be mistaken for isotropic. Test all high RI stones resting on a pavilion facet and not table-down.

Polycrystalline stones, such as chalcedony and jadeite, appear bright throughout the entire 360° rotation. This is because these stones are composed of tiny crystals that are oriented randomly; therefore some will always be in an inclined direction to vectorize light. Hydrogrossular garnet, a cryptocrystalline isometric stone, also appears bright throughout the 360° rotation, with the polarization generated by reflection from the random crystal grain surfaces.

Round faceted glass, or any isotropic stone such as spinel or garnet, frequently shows a unique effect, that of a Maltese cross – a dark cross that is wider at the outer edges against a lighter background. It remains throughout the rotation and is caused by polarization by reflection.

The polariscope in combination with a strongly converging lens can be used to resolve an *interference figure* to determine whether an anisotropic stone is uniaxial or biaxial. The instrument used in this way is referred to as a conoscope (alternatively, konoscope). The lens is usually a strain-free glass sphere, which is inserted between the test stone and the analyser. It may be a hand-held glass sphere on a rod, or be attached to the polariscope. To test a stone, manipulate it in the fingers between crossed polars until interference colours are seen as a rainbow effect. This occurs as the viewing direction approaches the optic axis. When the position is found where the colours are the widest and purest, position the glass sphere over the stone and view it through the analyser. This causes the light coming from the stone's optic axis to converge through the analyser and an interference figure appears. In general, one of three observations is made:

1) a shadowy four-armed cross appears in the centre of colour-fringed concentric rings. This is a uniaxial interference figure. The centre where the X crosses is the optic axis.
2) two arms can be seen. If there is a wide angle between the optic axes, then only one arm may be seen at a time as the other may be out of the field of view. This is a biaxial interference figure.
3) a bull's eye in which the arms of the typical uniaxial cross do not go through the centre of the cross. This is diagnostic for quartz.

With a cabochon or rounded bead the curved surface of the stone can act as the converging lens and it may only be necessary to manipulate the stone between crossed polars until the optic axis is vertical and the interference figure will appear floating on top of the curved surface. Also, if a glass sphere is not available, a drop of viscous fluid that will hold its domed shape, such as syrup, can serve as a condensing lens.

Anomalous extinction (AEX)

In certain instances it is possible to see an effect caused by strain that can mimic anisotropy in an isotropic stone. This is never detected on a refractometer but can be detected on a polariscope. It is commonly seen with glass, synthetic spinel, diamond and some garnets. When a mineral is truly birefringent, it appears successively and distinctly light/dark four times when rotated 360°. But when a stone with strain is rotated between crossed polars it shows vague patches or bands of shadow alternating with patches or bands of light. This is known as *anomalous extinction* (AEX), or sometimes by the misnomer *anomalous double refraction* (ADR). Anomalous double refraction is an unsuitable term because refraction is not the issue when considering extinction. Synthetic spinel is noted for its characteristic *tabby extinction*. With experience, these observations are easily distinguished from those of true anisotropy.

If in doubt there is a confirmation test for AEX. Find the position where the brightness through the stone is at its maximum on the dark background. Then rotate the analyser 90°. If the stone turns even brighter then AEX is the cause; if it remains the same or is reduced in brightness, then it is truly anisotropic. The confirmation test does not always work with garnets.

Colour

Colour is one of the primary contributors to the beauty of gemstones. It is purely a sensory perception. It is neither a property of objects nor of the light itself. Sir Isaac Newton was first to make the fundamental distinction between physical phenomena and perceptual phenomena and concluded on this basis that colour is exclusively a perceptual phenomenon. Newton observed that light waves are no more coloured than are radio waves; it is only that they have the power to stir up the sensation of colour. Thus, all colours are a purely subjective experience and the colour(s) we see in any object can be affected by the brightness of the light, the wavelengths of the incident light, and background colours.

It was seen earlier in the lesson on light that when a ray of white light is passed through a prism of glass the ray is fanned out into a rainbow of *spectral colours*, or a *spectrum*. Newton described the spectrum as consisting of the following colours: red, orange, yellow, green, blue and violet. Sunlight, or white light, is composed of all the colours from the visible portion of the electromagnetic spectrum ranging from violet (400nm) to red (700nm). Each particular wavelength designates a certain colour: 490nm corresponds to royal blue, 500nm to teal blue.

Humans can experience as white a light that consists of a mixture of just two properly chosen spectral components; any two spectral components which, when added together (as in additive mixing), appear as white are termed *complementary*. A mixture of one component that appears blue and one that appears yellow will look just like white. These complementary components cancel one another out, yielding a colourless sensation. Similarly, if an object absorbs light of predominantly one colour, then the remaining light will appear as the complementary colour. A stone absorbing the wavelengths corresponding to violet and blue will be seen as the complementary colour, yellow. Mixing components of light does not obliterate individual spectral components; these remain present, and their equivalent-to-white is a product of the visual system. Specialized instruments such as the spectroscope can distinguish these mixtures even though the human eye cannot.

In a series of experiments that selectively blocked or reduced light intensities, Newton was able to distinguish between pure light and light that is made up of several different components, termed *composite* light. Light of 589nm is pure yellow and cannot be broken down by a prism into constituent colours. Such light is termed *monochromatic*. This is not the only way in which a human with normal vision can perceive yellow. Sets of physically different distributions of light energy can produce identical colour experiences. Members of such sets are *metamers* and are said to be metameric. For example, an additive mixture of 530nm (green) light and 630nm (orange-red) light is metameric to and perceived as yellow.

In addition, by adding the appropriate components of light, colours unlike any of the pure lights could be produced. By subtractive mixing red and blue, Newton produced purple. Red pigment subtracts some wavelengths and blue pigment subtracts other wavelengths. Such colours are termed *non-spectral*. In rainbows one does not see colours such as purple and brown because non-spectral colours cannot be represented in the visual system by one wavelength; rainbow colours are all pure monochromatic light. Thus, from a perceptual standpoint, the visible spectrum has a gap; it does not contain all the colours that can be perceived by the eye.

In modern colour science, as with gemmology, the word 'colour' denotes three different qualities: *hue, brightness* and *saturation*. Hue refers to the quality that distinguishes among yellow, blue, green, etc. In everyday language we speak of the colours of the rainbow, but technically we are referring to the hues of the rainbow. Brightness, or lightness, is related to the amount of light. This is the same quality that allows one to designate a colour as 'dark green' or 'light blue'. Brightness is also commonly known as *tone*. The third colour quality, saturation, characterizes a colour as 'pale', 'vivid' or somewhere between these extremes. Saturation is related to the intensity of colour and allows one to designate a colour as 'vivid green' or 'pale orange'. The term *saturated* is applied to colours that are vivid and *desaturated* to colours that are greyish or washed out.

Selective absorption of light

The fundamental colour of a medium such as a gemstone is caused by the absorption of certain wavelengths of light, meaning certain colours, from the white light falling upon it, thus causing the residual colours, those that are transmitted or reflected and not absorbed, to combine and give colour to the gemstone. In opaque stones this absorption takes place at or near the surface and in transparent stones the absorption takes place when the light is passing through the stone. The perceived colour is *complementary* to the colours of the absorbed rays. Light passing through a ruby has the orange, yellow, green and violet absorbed from the spectrum: hence, the red and some of the blue are transmitted, which together give the typical colour of ruby. This effect is known as the *selective absorption of light*. Certain different absorption patterns may yield the same general hue, such as a green, but the greens may have slightly different brightness (light green, dark green) or intensities (vivid green, pale green). If a stone is colourless either small amounts of light are being absorbed or each wavelength being absorbed is accompanied by complementary wavelength absorption so the net result is white light.

Light absorption can be caused by many chemical elements, compounds and structures, the mechanisms of which are explained by the physics of ligand fields, charge transfer mechanisms, semi-conductor doping and colour centres, all of which are beyond the scope of this discussion. It is useful to note the distinction between actual physical causes of body colour and phenomena such as dispersion and interference, which contribute to the colours seen in gemstones. This lesson deals exclusively with primary body colour as caused by chemical elements and selective absorption, and how this may be exploited for analysis by the spectroscope.

In most gemstones colour is an accidental quality and, in general, gemstones are colourless when in the pure state. For example, pure corundum (aluminium oxide) is colourless white sapphire, but when containing a very small percentage of the oxide of the metal chromium it is red (ruby), and when the impurity is a combination of the oxides of iron and titanium the colour is blue (sapphire). When the colour-causing element is an impurity the gemstone is termed *allochromatic*. When the colour-causing element is an essential constituent of the chemical composition of the gemstone it is termed *idiochromatic*.

Idiochromatic stones such as peridot, turquoise and malachite are never colourless.

The selective absorption of light in both idiochromatic and allochromatic gemstones is mainly caused by the presence of one or more of the *transition elements*, alternatively known as the transition metals: elements 22 to 29 – titanium, vanadium, chromium, manganese, iron, cobalt, nickel and copper. These elements in small proportions have significant effects on optical properties, particularly colour and luminescence. They are characterized as being held together by relatively weak atomic forces, and their particles are able to vibrate freely when stimulated by any form of energy. When the form of energy is light, the vibration of the particles results in selective light absorption, producing a perception of colour in the object.

In certain instances, gemstones exhibit colour-related phenomena, such as colour change. The gemstone alexandrite is the most well known of those gems that display colour change as a result of metamerism, meaning that the colour changes depending on the light the stone is observed in. It is coloured by a transition metal impurity, chromium. Chromium colours are brighter than others because it yields clear-cut absorption bands and the colours that are not absorbed are left virtually at full intensity. Chromium absorption lines show clear maximum absorption in the yellow, green and violet regions and free transmission in the red. The fact that alexandrite has extreme transparency in the red, even though the mineral is green, contributes to its colour change in different lights. The red and green produced by chromium are almost evenly balanced. It is the position, intensity and breadth of the 580nm yellow-green band that determines whether the stone appears red or green. In light that is rich in shorter wavelengths, such as daylight and fluorescent light, the stone should be almost emerald green, and in light that is rich in longer wavelengths, incandescent light and candlelight, the stone appears rich red. This effect is also seen in some specimens of corundum, spinel and garnet.

Pleochroism

In any direction other than along an optic axis, anisotropic stones split light into two separate rays that are plane-polarized to vibrate at 90° to each other. In some of these stones there is a difference in the selective absorption of the two rays, meaning that a different portion of the visible spectrum is absorbed from each ray, causing the rays to emerge as different colours or as different shades of the same colour. The cause is termed *differential selective absorption*. The effect is called *pleochroism* and it is a valuable discrimination factor for gemstones. Pleochroism is eye-visible in andalusite, tourmaline and zoisite among others. In these stones it is a major contributor to their beauty and they are oriented to maximize this effect in the face-up when they are faceted. But generally, pleochroism is rarely seen by the naked eye, and in order to observe the colour difference an instrument is necessary. This instrument is the *dichroscope*.

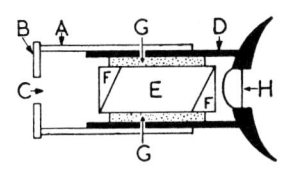

Figure 8.3 The dichroscope.

Dichroscope

The dichroscope is illustrated in Figure 8.3. It consists of a short metal tube (A) closed at one end by a metal end plate (B) that has a square aperture in the centre (C). Into the tube slides another metal tube (D) that contains a cleavage rhomb of optical quality calcite (E) to the ends of which are cemented glass prisms (F) that straighten out the path of light through the instrument. The calcite rhomb is held in place by packing (G). A low power lens is fitted at the end of the tube (H). Sliding the inner tube in or out focuses the instrument. Some instruments come with special stone holders.

On looking through the dichroscope at a light, two images of the square aperture (C) are seen side by side. These are two images of the same aperture with different vibration directions in each image: the light in one is plane polarized at 90° to the other due to the strong double refraction divergence of the calcite. To test a stone, place it close to the aperture, either on the stone holder or held in tongs, and view it through strong transmitted white light. Transmitted light is necessary because light can be partially plane-polarized by reflection and this can affect observations. The two plane-polarized rays of an anisotropic stone are separated by the strong double refraction divergence of the calcite and appear through the eyepiece side by side, one in each square for easy comparison.

If the two images of the aperture show identically the same shade and colour in all orientations and directions the stone is *monochroic*. It may be isotropic or an anisotropic – either uniaxial or biaxial – stone that does not show pleochroism. If the two images of the aperture show different colours, or different shades of the same colour, when tested in other than an optic axis direction, the stone is *pleochroic* and it *must* be anisotropic. The slightest difference in the colour of the two squares, or difference in shades of the same colour, determines the stone to be anisotropic. It is not the exact colours that form the test but whether or not a difference in colour exists. When pleochroism is seen, if two colours are produced the stone is said to be *dichroic*. The stone may be uniaxial, or a biaxial stone that shows only two colours. If three colours are seen the stone is said to be *trichroic*. Trichroism is only possible with biaxial stones, as it requires three different vibration directions. Note, however, that only two of the three colours can be seen at one time. The third colour is only observed by testing the stone in a different orientation and then it will appear in one of the squares, replacing one of the other two colours.

It is important to understand that pleochroism is the general term for the effect caused by differential selective absorption of light; it includes both dichroism and trichroism. The term 'dichroism' is often used interchangeably with 'pleochroism' to refer to the effect in general. This is incorrect. Dichroism should only refer to pleochroism of two colours.

Pleochroism is due to different absorption patterns for different vibration directions in anisotropic material therefore; it can never be seen in an isotropic stone. Pleochroism is only seen in coloured stones; it cannot be seen in colourless or white stones even if they are anisotropic. It is possible for pleochroism to be so faint that it cannot be observed even with the dichroscope. It is essential to remember that if an anisotropic stone is being viewed in the direction of an optic axis, pleochroism will not be seen even if the stone is strongly pleochroic because no anisotropic variation of light is seen in this direction. Also, pleochroism cannot be seen when the vibration directions of the stone and the calcite rhomb are at 45° to one another. Be certain to rotate the dichroscope around its length and to turn the stone and view it from several angles – at least more than three directions, as there can be two optic axes.

The dichroscope can readily distinguish anisotropic gemstones from isotropic gemstones, such as: (a) emerald from paste, composite soudé emerald and green garnet; (b) ruby from paste, red spinel and garnet; and (c) sapphire from paste or blue spinel.

Chelsea filter

The Chelsea filter provides a quick and convenient means of testing based on colour absorption. The small hand-held filter was originally produced to test for emerald and is still known as an *emerald filter* in the United States. It is constructed of coloured gel sandwiched between two layers of plastic and is designed to transmit only two areas of the visible colour spectrum: the yellow-green and the red, with virtually no transmission in the blue-green.

To view a stone through the Chelsea filter, hold the filter close to the eye with the stone about 20cm away in a strong light source – ideally, just under the edge of the shade on a bright light. This lights up the whole stone while the shade blocks stray light from distracting the eye. The light used is reflected, not transmitted. If it is difficult to see colour in the filter, the colour may be accentuated if the stone is viewed while touching a white background. Selective absorption occurs on the way through the stone and is reflected on the white background, and then occurs again when the light is on its way back out.

All chromium-coloured stones appear pink/red through the Chelsea filter. Most natural emeralds appear distinctly red, depending on the amount of chromium present: the finer coloured stones are vivid red and this ranges through to a pale pink for the lightest coloured emeralds. Green stones coloured with chromium are actually blue-green with a fair amount of red contributing to the colour. Most natural emerald has very little yellow-green in its colour and a great deal of red. The exceptions are some natural emeralds from India and South Africa. These do not show red because the iron in their composition quenches the effect of the chromium. Most green stones that imitate or can be mistaken for emerald appear green through the filter as they have little or no red in their colour, such as green glass, green doublets, and green tourmaline; however, chrome tourmaline, green grossular garnet and demantoid garnet appear red as they are coloured by chromium. Tourmaline can be easily differentiated from emerald with a refractometer reading and the latter two are over the limit of the refractometer, so they too can be easily distinguished. The Chelsea filter cannot differentiate between natural and synthetic emerald as they both appear red, but it is of note that synthetic emeralds typically show red of too great an intensity to be confused with naturals and this can cause suspicion.

Chrysoprase shows green through the filter because it is coloured by nickel and so it can be separated from dyed quartz, which appears red due to chromium. Keep in mind that a natural chrome chalcedony will also appear red, but this is a rare stone and it can be readily separated by spectrum: the natural has one band in the red and the dyed has three bands in the red. Jades dyed with chromium appear pink/red.

Materials coloured with cobalt also appear pink/red through the filter as these too transmit a significant amount of deep red light. The filter is most useful for detecting synthetic blue spinel and cobalt-coloured blue glass as these appear a rich red and can be distinguished from sapphire, blue zircon and aquamarine, all of which appear green to some degree. The exceptions are natural purple-blue sapphires from Burma. These appear red through the Chelsea filter due to a trace of chromic oxide. The only other natural blue stone that appears red is the rare natural cobalt-coloured spinel, although the shade of red is nowhere near as distinct as it appears with the synthetic in most cases.

The observation of the bright fluorescent red shown by natural Burmese rubies may be helpful in some cases. Synthetic rubies also appear red, but often with much greater intensity and this may be cause for suspicion in some cases.

In no instance is the Chelsea filter able to provide diagnostic information. All observations should be taken as indications only. Additional testing is always required.

Spectroscope

The fundamental colour of a gemstone, as mentioned previously, depends upon selective absorption: wavelengths are absorbed from the incident white light and the residual colours; (those that are not absorbed) pass through the stone and combine to produce the colour of the stone. Selective absorption is the basis for an important instrument known as the spectroscope, which analyses the light transmitted through or reflected from a gemstone. The spectroscope uses either an arrangement of prisms or a diffraction grating to produce a visible spectrum through the eyepiece. An ordinary incandescent light will give a *continuous spectrum*, with colours gradually changing from red at one end to violet at the other. The sun's spectrum is similar but is crossed by dark lines called *Fraunhofer lines*. Fraunhofer lines are caused by the selective absorption of light by various elements in our atmosphere and the sun's outer layer from the white light generated by the sun's interior. Because of these bands, sunlight is never used for spectral analysis.

When light transmitted through or reflected from a stone is examined by a spectroscope it will be seen that those parts of the spectrum totally absorbed appear as dark vertical lines or bands, and those not absorbed remain visible. This produces an *absorption spectrum*. Sometimes the energy from the light illuminating the stone is able to stimulate the element(s) such that they emit light rather than absorb it. This produces the fluorescent lines of an *emission spectrum*. The characteristic dark, or sometimes bright, vertical lines and bands crossing the range of spectral colours are an invaluable identification tool for many stones. The spectroscope provides the quickest and easiest identification of stones for which it gives diagnostic information.

Absorption spectra were first considered in relation to gemstones in a publication by A. H. Church in 1866, which gave notes on the narrow dark bands seen in the spectrum produced when white light has

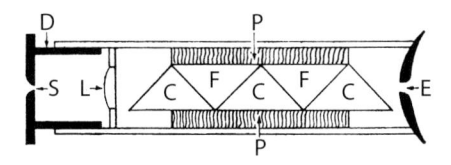

Figure 8.4 The parts of a prism spectroscope. S is the slit that admits light. L is a converging lens. E is the eyepiece. C is a prism of crown glass, which has lower dispersion. F is a prism of flint glass, which has high dispersion. P is packing, and D is the adjustable casing. The entire works are housed in a brass tube construction.

passed through a zircon. This was followed by the work of W. T. Wherry in the United States and B.W. Anderson and C. J. Payne in the United Kingdom who carried out research and published data of the absorption bands found in many gem materials. These absorption bands are in general due to some chemical element or compound – typically an impurity – in the stone, and which may or may not be the cause of colour. The transition elements produce characteristic absorption lines or bands. Often the same element causes lines or bands in different positions with different species, and this too is helpful in identification.

The most useful spectroscopes are the direct-vision hand-held instruments. The two versions are the prism spectroscope and the diffraction grating spectroscope. Each instrument is constructed with an adjustable slit. Figure 8.4 shows a direct-vision prism spectroscope. The slit that lets in the light is in the focal plane of the converging lens, which serves to render the rays parallel as they enter the prisms. The angle of dispersion from one prism is too narrow to allow examination of the spectrum, so a configuration of five, or alternatively three, prisms was developed to produce a broader spectrum without deviations. Crown glass is low dispersion and flint glass is high dispersion, and in this configuration the amount of refraction of the three low-dispersion crown glass prisms is equal to the amount of refraction from the two high-dispersion flint glass prisms. Those instruments with five prisms produce 10° of dispersion and those with three prisms produce 7° of dispersion. Some instruments come with additional optics for a calibrated wavelength scale that is superimposed on to the spectrum. The prism spectroscope provides a brighter spectrum, which many find easier to see, but the spectrum is compressed at the red end and spread out at the violet end. In the other instrument, the diffraction grating used to disperse light into a spectrum is usually a glass plate with a series of very fine, 3000–15 000 per inch, lines imprinted on it. The spectrum produced is not quite as bright, but the wavelengths are evenly distributed across the spectrum.

Before attempting any test, focus the instrument via the slit adjustment by pointing at sunlight and using the Fraunhofer lines or, alternatively, the emission lines from a fluorescent light. It is most effective having the slit narrower because the narrower the slit the sharper the absorption lines. There are two usual methods of observation with the spectroscope: transmitted light and reflected light. The best high-intensity source of white light is usually a fibre optic light with a flexible light guide that can be bent into position, although a penlight is often sufficient. To test a stone, bend the light guide in a 'U' shape such that the light is pointing upwards at the end and position the stone over the light. Whether this can be done in the hand or in tongs depends on the size of the stone. Put the spectroscope to the eye, move in close, to no more than 5–10cm away, and observe the spectra from a position parallel to the stone. To examine a stone in reflected light place it table-down on a dark background. Point the fibre optic light at the stone from one side and

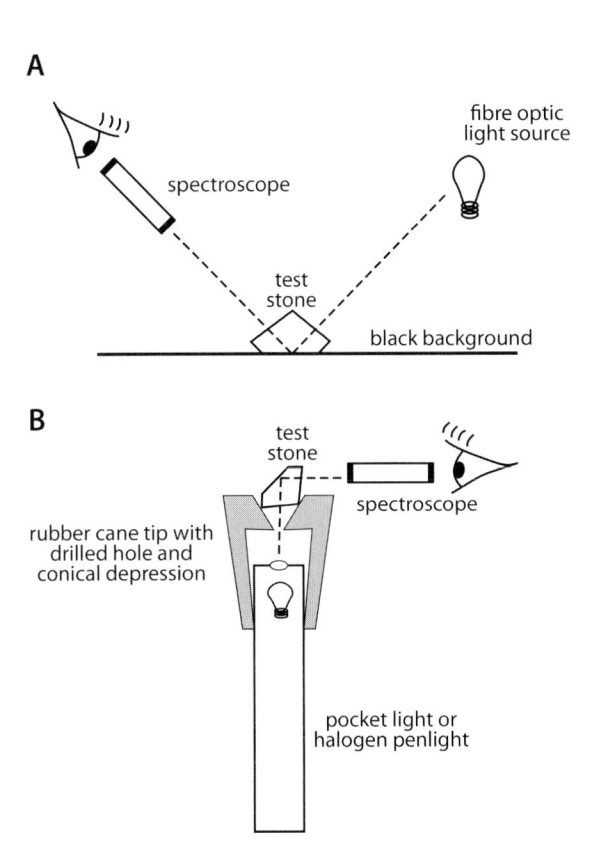

Figure 8.5 Two positions illustrating viewing a stone with the spectroscope. In the B position with the culet on one side of the hole and the girdle on the other side of the hole, a bright reflection off the internal surface of the table can be observed from the side through the spectroscope. An added advantage of the halogen penlight is that it can be focused to adjust the brightness of the light going into the stone.

allow the light to enter the stone's pavilion, reflect at the table, and exit from the pavilion on the opposite side. It may be necessary to turn a stone in several directions under different conditions to observe a spectrum, especially if there is pleochroism, or to determine there is no visible spectrum. For a subtle spectrum or pale-coloured stone, it can sometimes help to reduce the brightness of the light so that it does not overpower the lines or bands.

Spectroscopes are useful for testing faceted and rough gemstones. A darkened room absent of fluorescent light or sunlight makes focusing on the spectra easier and eliminates confusion from any bands caused by stray light. It helps to alter the eye's focus, as faint bands may be easier to see with peripheral vision. The acuity of the eye is less for the violet end of the spectrum than it is for the red end, so moving the head such that the red part of the spectrum is out of view makes it easier to see features in the violet. Passing the light through an appropriate filter or a flask of copper sulfate also eliminates the red wavelengths. Generally, when stones are heated they lose spectral features. Fibre optic light is a cool light, but if using another source do not overheat the stone in the light. It may even be helpful to cool it down prior to testing.

In certain cases the absorption spectrum can be diagnostic, meaning that if such an absorption spectrum is seen then the stone's identity can be confirmed with this information alone and further testing is unnecessary. The following commonly encountered spectra should be learned; additional spectra are available in the gemstone sections. It will be necessary to know information about the positioning of the bands and lines in theory, but overall absorption pattern – the relative strengths, widths and positions of the lines and bands – is most important for practical use. The commonly encountered spectra are:

Zircon – most zircons show a diagnostic spectrum of fine lines and narrow bands. Most show a 10–14 band spectrum, but those with up to 40 have been seen. The strongest line is in the red at 653.5nm, with prominent others at 660.5nm and 589.5nm. Other bands may be seen at: 691nm, 683nm, 662.5nm, 660.5nm, 653.5nm, 621nm, 615nm, 589.5nm, 562.5nm, 537.5nm, 516nm, 484nm, 460nm and 432.7nm. Metamict and heat-treated may only show the line at 653.5nm and it may be faint. These absorption bands are due to a trace of uranium and are thicker and more pronounced than the 'fine line' spectra of rare earth elements.

Ruby – a strong doublet in the deep red at 694.2nm and 692.8nm is usually seen as a single bright fluorescent line. Two fainter lines in the orange at 668nm and 659.5nm may also appear as fluorescent lines. A strong general absorption of yellow and green is centred on 550nm. There is transmission in the blue with three narrow bands at 476.5nm, 475nm and 468.5nm and a strong absorption of the violet. The spectrum, due to chromium, is diagnostic: any red stone with a bright fluorescence line in the red and narrow lines in the blue is, for certain, a ruby. It is easier to see the fluorescent spectrum if not looking directly at the light through the stone – just look at the general body colour.

Sapphire – spectrum is due to iron and is common to blue, green and yellow sapphires. Three bands in the blue, the strongest is at 450nm and this one may coalesce with the band at 460nm if the spectrum is really strong. The other band at 471nm is separate. The spectrum is diagnostic. In blue, green, orange and yellow stones the bands may vary in strength depending on the origin of the stone. Stones from Australia show all three bands strongly, but Sri Lankan stones show only a faint band at 450nm, as there is little iron content. Yellows from Sri Lanka exhibit no spectrum.

Yellow, Greenish-yellow and Brown Chrysoberyl – strong broad iron band centred on 444nm. Weaker narrower bands at 485nm and 505nm in the blue-green. Diagnostic.

Almandine Garnet – three strong broad bands centred in the yellow at 576nm, green at 526nm and blue-green at 505nm. Weaker broad bands in the orange at 617nm and blue at 462nm. Diagnostic. In very dark stones the 526nm and 505nm may look like one band. Almost all pyropes show the band at 505nm, as do many Spessartite and Sri Lankan garnets. Due to ferrous iron.

Dark Blue Cobalt Glass – three broad bands centred on 655nm, 590nm and 535nm. The centre band is the narrowest of the three. The exact positions vary with the composition of the glass. Diagnostic.

Synthetic Blue Spinel – shows strong transmission in the blue and red. And a cobalt spectrum of three broad bands centred at 622nm in the orange, 590nm in the yellow, and 544nm in the green. The centre band is the widest of the three broad bands.

Blue Spinel – a broad band in the blue centred at 459nm, a narrow band of equal strength at 480nm, a moderately strong band at 630nm and other weaker bands at 433nm, 443nm, 508nm, 555nm and 585nm. Due to ferrous iron.

Emerald – coloured by chromium it shows fine lines in the red, weak diffuse central absorption from 580nm to 625nm, weak lines in the blue at 477.5nm and 472.5nm, and absorption of the violet from 460nm. The lines in the red are sharp and dark. The most prominent are 683nm and 680nm; this is a doublet. Other lines may appear at 662nm, 646nm and 637nm.

Red Spinel – broad absorption in the green, absorption of the violet, and no lines in the blue. A characteristic group composed of closely spaced lines in the red centred on 680nm, which are often fluorescent. The emission lines in the red, caused by chromium, are striking and described as 'organ pipes'. The two brightest lines are at 686nm and 675nm. There are three weak lines on the long wave side and five on the short wave side of the dominant pair. The organ pipe of natural red spinel is diagnostic. It separates it from synthetics. It is one of the few cases where it is not necessary to look at inclusions through the microscope.

Demantoid Garnet – coloured by chromium with a doublet in the deep red centred on 701nm, a weak sharp line at 693nm, and bands at 640nm and 621nm. Some specimens do not show these bands due to general absorption in the red. At 443nm there is a strong band due to iron, which owing to the general absorption of the violet may appear as a sharp cut-off when no violet is seen beyond the band.

Peridot – three bands in the blue due to iron. One centred on 493nm, a fairly narrow band at 473nm, and a broader band at 453nm.

Yellow Apatite – a group of about seven fine lines in the yellow with the most visible at 584nm and 578nm and a smaller less prominent group of about five lines in the green centred on 520nm. These are due to rare earth didymium. May be seen very weakly in all calcium minerals.

Synthetic Sapphire – blue, green and orange synthetic sapphires never show the strong iron band at 450nm. If a suspected stone is found to show this band in the blue-violet, it is definitely natural; however, absence of this band is not proof-positive of synthetic origin as some naturals show it weakly or not at all. Colour-change sapphire imitating alexandrite does not show any bands in the red as the natural does.

Diamond – Cape Series diamonds have a signature line at 415nm caused by the aggregated nitrogen impurity. As the absorption at 415nm becomes stronger, so do the yellow body colour and the appearance of other Cape Series bands at 423nm, 435nm, 452nm, 465nm and 478nm. The spectrum is characterized by almost no absorption in the red. Diamonds from the brown series show a prominent band in the blue-green as well as weaker bands in the blue-green.

LESSON 9

Luminescence, Electrical and Thermal Properties, X-rays and Magnetism

Luminescence is the process by which certain materials are excited and emit visible light without heat; the material glows while remaining at low or room temperature. In luminescent effects, one form of energy is changed into a form of energy that can be seen. It is sometimes known as *cold light* as it is in opposition to incandescence, which occurs when something is heated in order to make it glow – in this case, heat energy is turned to light.

The production of neon lamps from electronically excited gases is a luminescent effect, as is the glow of fireflies caused by certain inorganic materials undergoing oxidization. In 1852 English mathematician and physicist George Stokes correctly identified this light as a *luminescent emission*. For emissions that ceased immediately when the incident light was removed, Stokes proposed the term *fluorescence*, named after fluorspar. Three decades later, German physicist Eilhard Wiedemann coined the term *luminescence* from the Latin to refer to all phenomena in which light is emitted without a rise in temperature.

It is important to begin to use this terminology correctly. Many sources use the term 'fluorescence' indiscriminately to refer to the various phenomena in general, but this is incorrect. Fluorescence is a specific type of luminescent effect and its usage should be restricted to denote short-duration luminescent effects occurring only while the material is in the presence of the cause of excitation.

Types of luminescence

There are numerous specific phenomena covered by the term 'luminescence' and these are given names employing a prefix indicating the means of excitation. The most important to gemmology is *photoluminescence*, which occurs when the glow is induced by visible light: electromagnetic radiations such as ultraviolet light and sunlight. There are two types: *fluorescence*, which remains only while the exciting radiation is incident on the material; and *phosphorescence*, which persists for an appreciable time after removal of the exciting radiation. Röntgenoluminescence is produced in response to excitation by X-rays. The prefix is derived from the name of the discoverer of X-rays, Wilhelm Röntgen. X-rays are found in the electromagnetic spectrum and, for this reason, röntgenoluminescence is not usually referred to by its proper name; it is more often simply termed X-ray fluorescence and thought of as a division of photoluminescence.

There are also: *chemiluminescence* – produced as a result of chemical change as shown by the glow of slowly oxidizing phosphorus; *bioluminescence* – produced as a result of chemical decay in living things as shown by the glow of fireflies. This decay involves chemical change, making it a sub-class of chemiluminescence; *triboluminescence* – produced as a result of mechanical disturbance such as breaking, rubbing or scratching; *thermoluminescence* – released from materials that are warmed. This is a very different concept than heat producing light by means of incandescence. The heat merely accelerates emission of light by a material that has previously been excited by some other means. The heat is strictly the release mechanism for delayed phosphorescence, it is not the cause; *electroluminescence* is produced as a result of excitation provided by electrons; and *cathodoluminescence* is produced in response to cathode rays. Since cathode rays are high-velocity electrons, it is a sub-class of electroluminescence. Diamond displays the greatest variety of luminescent effects, showing all but chemiluminescence and bioluminescence.

Crossed filters

What is luminescence? In 1852 Stokes used a 'crossed filter' technique to prove his assertion that a material's fluorescence is always of a longer wavelength than that of the original excitation. B. W. Anderson later applied the technique to gem testing. At the outset, understand that this technique employs coloured filters and should not be confused with the 'crossed polars' of the polariscope. The crossed filter method utilizes a strong white light, a blue filter and a red filter. The blue filter – at that time it was a flask containing a saturated solution of copper sulphate – is placed before the white light. Typically this is arranged on a black background. The filter passes blue and violet light and absorbs the red, orange, yellow and most of the green. The flask acts as both a filter and a condensing lens for the light. At present, it is easier to use a simple commercially available gelatine filter sandwiched between glass protectors. With the naked eye there appears a patch of blue light on the black background. Next another filter is held in front of the eye that passes only red light. This obliterates the patch of blue light and no light is seen. The test stone is placed on the black background such that the blue light illuminates it and no other stray light reaches it. When the red filter is held to the eye, any stone that appears to glow red must be fluorescing. It must be stimulated by the energy in the blue light and be re-emitting part of that energy as red light – light of a longer wavelength – since the light illuminating the stone contains no red component.

Red fluorescence is characteristic of minerals coloured by chromium. Ruby, red spinel, alexandrite, emerald and pink topaz will all show red fluorescence between crossed filters. But this is diminished in proportion to any iron present and the synthetics coloured by chromium also show the same effect, although often stronger. The crossed filter test is useful to distinguish between chromium-coloured emerald and green beryl coloured by vanadium, and also alexandrite from chrysoberyl.

A practical elaboration is to use a spectroscope in place of the red filter. In this instance ruby can be differentiated from spinel, as ruby (and synthetic ruby) will show the typical sharp bright red line due to the doublet at 692.8nm and 694.2nm, with only faint fluorescent lines on either side. On the other hand, natural red spinels will show the typical 'organ pipe' spectrum of fluorescent lines in the red: a group of several narrow lines with two of equal strength at 686nm and 675nm in the middle. It must be noted, however, that synthetic red spinels show a single line at 686nm surrounded by fainter lines, and thus it more resembles that of ruby.

The mechanism of luminescence

The excitation of luminescent material is typically accomplished by an injection of energy provided by ultraviolet light, X-rays, electrons, or alpha particles. The resulting colour of the luminescence is principally determined by the chemical composition of the material. The resulting intensity of the luminescence depends on: (a) the chemical composition of the material; and (b) the intensity and nature of the excitation. The essential action is electronic. The process involves the localized absorption of energy that excites one or more electrons bound to an atom, or a group of atoms, without requiring agitation of the atoms as a whole. The excited electron emits at least part of its excitation as light energy.

In contrast, light emission by incandescence, thermal radiation, is the result of simply heating the material until all of the atoms are strongly agitated. The colour and intensity of resulting light emission in this case depends primarily on the temperature of the material. Most emission is at the far infrared end of the spectrum, but as temperature increases, it shifts to shorter wavelengths – orange, yellow and white. The hotter it gets, the shorter the wavelengths produced. When the wavelength is in the visible portion of the spectrum, incandescence is produced. Very hot tungsten filaments are responsible for the light from ordinary incandescent light bulbs.

The basic mechanism of luminescence is a simple concept: an electron(s) absorbs energy, becoming electrically excited, and then returns to the unexcited state while emitting some of the absorbed energy as a photon of light. Thus, luminescence is a process in which one kind of energy is changed into a visible kind of energy. The detailed mechanism of luminescence is complex because sub-atomic particles have delicate and complicated interactions, but a simple explanation based on quantum electrodynamics is suitable for gemmological purposes. Quantum theory postulates that waves of the electromagnetic spectrum are of a particle nature, these particles being quanta or photons. Atoms only absorb or emit a set definite quantity of light, that being one photon of a given wavelength.

The electrons orbiting the atomic nucleus in the outer shell play the important role. For each electron shell, there are different energy levels at which electrons can reside. Atoms only exist in certain states: the lowest state is termed the *ground state*, and specific higher states are termed *excited states*. If a photon combines with an electron, it adds energy to the electron, allowing the electron to move up to a higher energy level within the same shell. Thus, it is raised from ground state, its inner stable orbit around the nucleus, to an excited state. In most cases of light travelling through a transparent medium, the photon that enters the medium is not the one that makes the final exit; one photon joins the electron and becomes part of the energy of that electron in the excited state. When the electron returns to the ground state it gives up the extra energy as a new photon with properties identical to the original; the new photon can then join another electron that later produces a subsequent new photon, and so on.

If the transition to the higher energy level and return to ground state is fast and one single quantum is released, then simple transmission or remission of light is observed. If the return to the ground state is fast and more than one quantum (at least one of which is visible) is released, the emission of fluorescence results. Fluorescence represents the spontaneous return of the electron to the ground state with a change in the wavelength of light, the balance of energy usually including a phonon of heat.

If the electron is excited to a *metastable* level, the material will glow during the period of excitation and also, after the source is removed when the electron finally releases the photon, phosphorescence results. In some instances, electrons can be excited to a really metastable level at room temperature, one in which they are 'trapped', and in this instance heat is required to make the electron release the photon. The glow produced in this case is thermoluminescence, which can be thought of as frozen-in fluorescence. It is important to grasp the concept that the initial excitation is the cause of this glow; the heat, which is always less than necessary to produce incandescence, is like a key to unlock the energy and release the glow. If the material was not subjected to the excitation in the first place, the addition of heat would not produce a glow. Heat causes the electron to give up the photon by kicking it back up from its trapped state (between the ground state and higher excitation states) to a higher level of excitation; one from which it can return to ground state and release the photon. Thus, whether the material phosphoresces or thermoluminesces depends on the degree of stability of the metastable level. The difference between fluorescence, phosphorescence and thermoluminescence is how long it takes the electron to release the photon: if the electron releases it right away, fluorescence is produced; a longer release time produces phosphorescence; and after a longer period and the requirement of heat, thermoluminescence is produced. If a material phosphoresces, it must also fluoresce. Likewise, if a material thermoluminesces, it must also phosphoresce and fluoresce. And, with each of these different types of luminescence the colour and intensity of a material's response may vary.

It is common to find gem species, such as diamond, in which only some members of the species exhibit luminescence. Furthermore, it is affected by impurities. When present, manganese is thought to promote fluorescence, and iron is thought to quench it.

Testing luminescence

In gemmology the most important luminescent responses are those of ultraviolet light. In the electromagnetic spectrum, ultraviolet (UV) light covers wave bands from just below 400nm down to 10nm, but the usable range is between 200nm and 400nm. Long wave ultraviolet light (LWUV) has a principal emission at 365nm. Short wave ultraviolet light (SWUV) has a principal emission of 253.7nm. Mercury vapour lamps are used to produce ultraviolet light. LWUV is produced by a high-pressure mercury discharge lamp with a suitable filter, usually Wood's glass, which is a cobalt glass with about 4% nickel oxide. SWUV is produced by a low-pressure mercury lamp with an OX7 filter. For gemmological purposes convenient combination lamps with both LW and SW light sources are marketed and these allow for rapid comparisons between LW and SW responses. Stones to be tested should be placed a short distance below the lamp with the light directed on to the stone. Luminescent effects are best seen in a darkened room, or alternatively in a viewing cabinet. It is never necessary to look directly at the light, only at the stone under the light; however, it is worth mentioning that UV light, in particular SWUV, is harmful to the eyes so as a precaution, if necessary, wear protective glasses – even those of ordinary glass are sufficient to filter the SWUV. With stones of weaker response it may be necessary to pick them up in tongs and hold them as close as possible to the light. Limit exposure during testing as the colour of some stones, such as heat-treated blue zircon, can fade if exposed for long durations. If this occurs, the colour can be restored if the stone is heated to a dull red temperature.

Table 9.1 LWUV and SWUV fluorescence

Listed are the typical responses for various gemstones.

Gemstone	LWUV	SWUV
Apatite – yellow	Lilac	Lilac/pink
Apatite – blue	Light to dark blue	Blue
Apatite – green	Mustard yellow	Weak yellow
Apatite – violet	Greenish-yellow	Pale mauve
Alexandrite	Weak red	Weaker red
Amethyst	Inert	Blue (some)
Aquamarine	Inert	Inert
Benitoite	Inert or dull red	Bright blue
Chrysoberyl	Inert	Inert
Cubic zirconia	Inert or faint yellow	Faint yellow
Diamond – Cape Series	Blue*	Weaker blue*
Emerald	Red	Red
Fluorite	Violet/blue	Weaker violet/blue
Garnet	Inert	Inert
Kunzite	Orange	Weaker orange
Ruby	Crimson red	Crimson red
Sapphire – pink	Red	Red
Sapphire – blue	Inert	Inert
Sapphire – yellow Sri Lankan	Apricot-yellow	Apricot-yellow
Scheelite	Inert	Bright whitish-blue
Spinel – red	Red	Weaker red
Spinel – dark blue	Inert	Inert
Spinel – pale blue	Green (some)	Inert
Synthetic moissanite	Inert	Inert
Topaz – blue and colourless	Weak yellowish	Very weak yellowish
Tourmaline – red and pink	Inert	Blue to violet
Tourmaline – yellow, brown and green	Inert	Strong yellow (some)

*Diamond can fluoresce any colour to any degree or not at all. Blue is the typical Cape Series response

Luminescent responses

Not all gemstones exhibit luminescence and most exhibit a range of colours and strengths, but in a number of cases the responses to LWUV and SWUV can assist in identification. Luminescence testing is a confirmation test and not diagnostic in itself, with one exception. The only instance in which a luminescent response is diagnostic: blue fluorescence followed by yellow phosphorescence is proof positive of diamond. Typical responses are listed in Table 9.1 and notes are given for special consideration below.

Diamond

Diamond's UV response is not always uniform, it can be zoned or patchy, but the SW response is generally less intense than the LW response. Response to LWUV and SWUV can produce various colours and intensities. Cape Series diamonds usually glow blue to violet, and combined with yellow phosphorescence this is a diagnostic result. Brown series diamonds typically glow greenish. Diamonds that photo-luminesce to visible light energy in the form of sunlight usually display a blue or green colour. Some colourless diamonds, termed Jagers, display a strong blue response. Most natural green diamonds owe their colour, at least in part, to the emission of a green luminescence that overpowers a yellow or brown body colour. The passing of an electrical current through diamond induces an electroluminescent response in some stones. The colour is usually blue-green and the response is far more likely in Type II diamonds. Triboluminescence may be exhibited when diamond is rubbed on certain materials. The strongest glow is produced across the grain of wood. Diamonds may also glow a dull red in response to the polishing process. Thermoluminescent responses have not been observed in diamonds of natural origin; however, they have been produced in a test sample of near colourless synthetic diamonds immersed in hot water subsequent to excitation by UV radiation. Cathodoluminescence has been employed as a research tool to explore diamond origin and properties, including the structure of platelets.

Corundum

Rubies (natural and synthetic) show a strong crimson glow with LWUV and SWUV due to chromium but synthetic rubies typically show a much brighter glow and this may be a helpful observation. Violet and pink sapphires also show red under UV. Natural blue sapphires are inert due to iron content, but synthetic blue sapphires show green or sometimes blue under SWUV. Natural green sapphires are inert, but synthetic green sapphires glow red under LWUV. Yellow sapphires are variable: iron-rich yellows are inert, but the Sri Lankan is apricot-yellow under UV and this can distinguish it from from the synthetic. This is useful because yellow synthetics typically have no curved growth lines or bubbles.

Beryl

Natural emeralds usually show red to UV and the synthetics typically a stronger red. Aquamarines and other beryls are generally inert.

Spinel

Red and pink, natural and synthetic spinels show a red glow under UV, like that of ruby. Natural dark-blue spinels are inert due to iron content, but pale blue sometimes exhibit green to LWUV but are inert to SWUV. Synthetic colourless spinels are inert under LWUV and bluish white under SWUV. Synthetic blue spinel, made to imitate blue sapphire, shows red under LWUV, and red, orange or bluish-white under SWUV.

Other stones

Fluorite generally glows blue or violet, stronger under LW than SW, but the Blue John variety is inert under both. Kunzite glows orange under LWUV and is inert to SWUV. Benitoite glows bright-blue under SWUV, but is inert or dull red under LWUV. This gives a distinction from natural sapphire. Sheelite fluoresces bright whitish-blue under SWUV, but is inert under LWUV.

LUMINESCENCE, ELECTRICAL AND THERMAL PROPERTIES

Electroconductivity

The property of transmitting electricity is common to metals; they transmit large electrical currents under the influence of a small difference of potential. The reciprocal of conductivity is resistivity, a measure of how strongly a material opposes the flow of electric current. Certain gemstones will conduct electricity when inserted into a suitable electrical circuit – these include hematite and synthetic rutile; however, only in the case of diamond is electroconductivity useful as a test.

Semiconductors are so named because they are roughly intermediate in electrical properties between metals and insulators. Type IIb diamonds, including all natural blue diamonds, are electrical semiconductors – the only natural diamonds that conduct electricity. When found in a near-colourless diamond, electrical conductivity is a strong indicator of synthetic origin. Electrical conductivity is most important for the identification of natural blue diamond as it can be reliably separated from non-conducting artificially coloured blue diamonds on this basis.

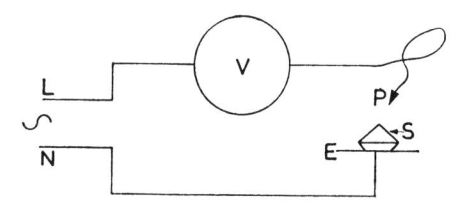

Figure 9.1 A simple circuit for testing a stone for electroconductivity: L is the live side of the alternating current of the mains circuit and N is the neutral side. Batteries may be used in place of the mains. V is a suitable voltmeter. E is the electrode upon which the test stone is to be rested. P is the probe to make contact with the stone.

Photoconductivity

Photoconductors are materials that exhibit high electrical resistance in the dark, but show a marked increase in conductivity on exposure to light. It was on the basis of differences in luminescence and photoconductivity that Type II diamonds were originally separated into subgroups IIa and IIb. The photoconductivity of Type II diamond is good and, in particular, Type IIb are photoconductive to gamma rays, meaning that when gamma rays fall on a IIb diamond, the current of electricity passing through the diamond is amplified. For this reason, Type IIb diamonds are applied in the field of radiation detectors.

Piezoelectric effect

The piezoelectric effect refers to the ability of certain crystals to become electrically charged when subjected to mechanical stress. The effect is found to occur in hemihedral crystals, the most important of which are quartz and tourmaline, although it is also displayed by topaz. In a piezoelectric crystal the positive and negative charges are separated, but are evenly distributed so that the crystal has a neutral charge overall. When stress is applied it disrupts this symmetry and the asymmetry of the charge generates the voltage. When quartz is cut along a certain axis and compressed it develops a positive and negative charge on opposite faces. When the same quartz is placed under tension, meaning that it is pulled or stretched, the polarity of the electrical charge reverses.

What makes the effect most interesting is that all piezoelectric materials possess the converse effect in which the application of an electrical field creates mechanical stress, in the form of distortion, in the crystal. The distortion is due to the fact that the charges inside the crystal are separated so the applied voltage affects different points in the crystal in different ways, and the result is distortion. The length diminishes and the thickness increases when the electric field is in one direction, and the thickness diminishes and the length increases when the field is applied in the opposite direction. Thus, a quartz plate could be made to oscillate if the applied current was alternating. This effect means that quartz crystal oscillators can be used in electronic devices such as those for the control of accurate clocks. The crystal must be untwinned and synthetic quartz is produced in quantity for these applications.

Pyroelectricity

Stones that exhibit pyroelectricity develop an electrical charge across opposite ends of their crystal axis when they are heated. This occurs in tourmaline with an opposite charge at opposite ends of the crystal, and in quartz with the opposite charge at alternate corners of the crystal.

This effect is the cause of such stones readily collecting dust when heated by sunlight in a shop window or by the bright lights in a jewellery display case.

Triboelectric effect

Triboelectric materials develop an electrostatic charge when subjected to contact and separation with another material, as in rubbing. This is static electricity. Amber is most noted for displaying *triboelectricity* when it is rubbed briskly with cloth, wool in particular. It becomes capable of picking up small particles such as fragments of tissue paper. It is not useful for positive identification since a number of amber imitations possess the same ability; however, if no frictional electricity is observed then it cannot be amber.

Thermal conductivity

Thermal conductivity refers to the ability of a substance to conduct heat. Almost all gemstones are poor conductors of heat; the one exception is diamond. All Types of diamond are very good heat conductors, a function of their atomic structure: the lighter the atoms, the easier they are moved. In metals, free electrons conduct heat. In the diamond structure, energy produces heat by stimulating atoms into vibration and heat is conducted by lattice vibrations, called phonons – individual heat corpuscles. The rigidity and relative freedom from flaws make the diamond structure efficient at transmitting vibrational energy. The phenomenal thermal conduction of Type IIa diamond is owed to its freedom from impurities.

Thermal conductivity is expressed in Watts per metre per degree Celsius. For Type I diamond at room temperature it is 1000 W/m/°C, over two times more conductive than copper (385–401W/m/°C). Type IIb diamond is 1600 W/m/°C, and Type IIa diamond has the highest thermal conductivity, 2600 W/m/°C, 6 times more conductive than copper at room temperature. This high thermal conductivity is evidenced by the fact that diamond feels cool to the touch – the property behind the nickname 'ice'. Touching diamond and a simulant to the tongue will reveal that only the diamond feels cold, as it readily absorbs heat from the tongue, which makes it feel cold. Electronic devices not much larger than a pen have been devised to exploit diamond's high thermal conductivity and serve to separate it from simulants on the basis of this property alone. The only simulant close to diamond in thermal conductivity is synthetic moissanite with a conductivity of 200–500 W/m/°C; the next closest is corundum at 40 W/m/°C.

X-rays

X-rays, discovered by Wilhelm Röntgen in 1895, comprise part of the electromagnetic spectrum with wavelengths approximately 10 000 times shorter than visible light, with a mean wavelength of 0.10nm. X-rays are generated in a vacuum tube, termed an X-ray tube or a Coolidge tube, using high voltage to accelerate electrons released by a hot cathode, termed a filament – a negative pole – directed towards a heavy metal target known as the anode – a positive pole. The electrons striking the metal target, typically tungsten, dissipate most energy as heat and a smaller proportion as X-rays. The wavelength ranges of X-rays and gamma rays overlap and their radiations are defined by the source: X-rays are emitted by electrons outside the nucleus and gamma rays are emitted from the nucleus of radioactive materials. X-ray machines are large, costly and require many safety precautions so their usage is typically restricted to a laboratory setting.

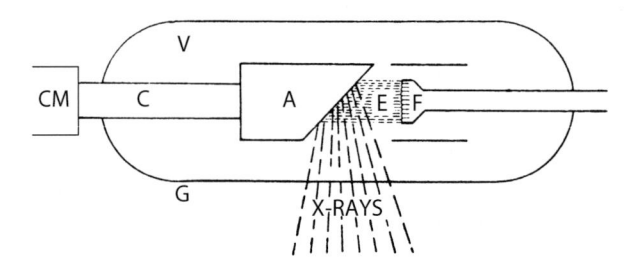

Figure 9.2 Schematic diagram of an X-ray tube, known alternatively as a Coolidge tube. G is the glass envelope. V is the vacuum that exists inside the glass envelope. F is the hot cathode, or filament. E is the stream of electrons. A is the anode or tungsten metal target. C is a copper stem and CM is the cooling mechanism.

X-rays have played an important part in developing the study of crystallography. In the early twentieth century it was discovered that a beam of X-rays is scattered when it strikes a crystal in the same manner as a beam of light is scattered when it strikes a ruled grating. A technique, which came to be known as X-ray diffraction, made it possible to determine the molecular structure of solids. In 1912 Max von Laue discovered that radiations in the X-ray wavelength range of 100–1nm are diffracted by crystals whose layers of atoms are only a few nanometres apart. The crystal's atomic lattice planes reflect the X-rays and form *Laue spots* on the diffraction pattern. Eventually, diffraction methods progressed to where it was possible to reflect X-ray beams off the crystallographic planes parallel to the crystal surface and off those that are perpendicular to the crystal surface. It was revealed that the planes in a crystal were members of a large network of planes, all of the same orientation, all equidistant and parallel, and each studded with atoms in the same pattern and density. Thus, the nature of crystals as three-dimensional latticework patterns of atoms was revealed and the structure of crystals could be studied at the atomic level.

Since X-rays penetrate solid objects they are useful in gem testing to provide information about the internal nature of gemstones. When X-rays interact with a substance or material, some are absorbed, some are scattered, and some pass through. The relative amount of each outcome depends upon the length of the X-rays and the structure and composition of the substance. Since electrons absorb X-rays, it stands to reason that a material that is transparent to X-rays would be composed of atoms that have few electrons. This is the principle behind the application of diagnostic radiography in medicine. Bones have a higher electron density than soft tissue, allowing the X-rays to pass through flesh and leave a white shadow image of the bones on the photographic film. Since diamond is carbon, atomic number six, with only six electrons per atom it is highly transparent to X-rays. The general relationship is that opacity to X-rays increases somewhere on the order of the fourth power of the atomic number, which is roughly half an element's atomic weight. X-ray transparency is the single best diagnostic test for identifying diamond: diamond is completely transparent to X-rays and all other gems and simulants are opaque to some degree in proportion to their atomic composition. Cubic zirconia is completely opaque as the zirconium atom has a mass of 91.22 and readily absorbs X-rays. It is worth noting that transparency to visible light is irrelevant; opaque industrial diamonds, because they are still carbon, are transparent to X-rays. It is a quick test, requiring only 5–10 seconds of exposure, and can be done with a mounted stone, so it is an extremely useful test when the circumstances require it.

Table 9.2 X-ray fluorescence

Listed are the typical responses for various gemstones.

Apatite – violet	Bright yellowish-green
Alexandrite	Very dim red or inert
Benitoite	Blue
Calcite	Orange
Chrysoberyl	Inert
Cubic zirconia	White
Diamond	Bluish-white
Emerald – natural	Red
Emerald – synthetic	Strong red
Fluorite	Blue or violet
Kunzite	Strong orange
Ruby – natural	Red
Ruby – synthetic	Red
Sapphire – blue Sri Lankan	Pink/red
Sapphire – synthetic blue	Green/blue
Sapphire – pink	Red
Sapphire – synthetic pink	Red
Scheelite	Blue
Spinel (red, natural and synthetic)	Red
Spinel (dark blue)	Inert
Synthetic moissanite	Inert
Strontium titanate	Inert
Topaz – blue and colourless	Greenish-white/violet-blue

Like UV light, X-rays can induce a visible glow of light in some substances and X-ray fluorescence can be of diagnostic value. Observation of X-ray fluorescence is what forms the basis of the X-ray separator, a machine used in many diamond mines to differentiate diamonds from waste rock and other minerals. Almost all diamonds fluoresce bluish-white to X-rays. It is a reliable and diagnostic response. Diamond simulants are typically non-fluorescent to X-rays, and those that do respond do not show the same colour range. With rubies, both synthetics and naturals respond red to X-rays, but the Verneuil synthetics exhibit distinct phosphorescence. X-ray fluorescence is presented in Table 9.2.

Magnetism

Magnetism is defined by the degree of attraction to a magnetized body: (1) ferromagnetism – materials characterized by a strong attraction, a property possessed by iron and some other metals; (2) paramagnetism – materials characterized by a less strong attraction; and (3) diamagnetism – materials repelled by a magnet. In general, magnetism is of little use in practical gem testing. Type Ib diamond is paramagnetic as it contains isolated substitutional nitrogen: one nitrogen atom substituting for one carbon atom. This results in an unpaired electron that makes it active in electron paramagnetic response (EPR). There are some other gemstones that are paramagnetic, and these include almandine garnet, spessartine garnet, rhodochrosite, rhodonite and hematite, but they are never differentiated on this basis. Investigations into magnetism are purely for scientific experiment with one exception: synthetic diamond.

Iron, nickel and cobalt are ferromagnetic. They can be permanently magnetized with an exceptionally high energy upon application of an external magnetic field, making them, in effect, tiny powerful magnets. Since these ferromagnetic metals are the most favoured metal catalysts for synthetic diamond production the residuals of the catalysts that remain in the diamonds cause them to be magnetic. Synthetic diamonds, even if they have no visible magnetic inclusions, have sufficient residuals to be attracted by a magnet. Natural diamonds are never attracted to a magnet, so testing for magnetism can be diagnostic. While an absence of magnetism is not proof-positive of a natural diamond, the presence of magnetism is proof-positive of synthetic origin. There are several simple methods by which this may be done, employing the rare earth 'magnetic wand' developed by Alan Hodgkinson or the Linton Synthetic Diamond

Sensor. One method useful for larger diamonds is to suspend the diamond on a nylon or cotton thread. Ensure the diamond is perfectly still and not in the path of any airflow that will cause it to move. Draw the magnet close to the crown of the diamond and observe any movement towards the magnet.

LESSON 10

Loupe and Microscope

Before examination of any stone can take place with the loupe or microscope, the stone must be cleaned to remove dirt, grease and foreign materials that can be confused with features in or on the stone. Stones can be cleaned by dipping them in a mixture of a few drops of dishwashing detergent in a bowl of water. The stone should be wiped dry in a lint-free cloth. Alternatively, solvents such as alcohol may be used, but if there is any doubt that the stone may be damaged by alcohol, it should not be attempted. Alcohol readily damages organics such as coral, oiled emeralds and plastic impregnated stones.

When the stone being examined is diamond, either for identification, grading or origin, it should be washed in alcohol, acetone or a solution of dishwashing detergent and cleaned in a special cloth. This is a lint-free nylon diamond cloth. The cloth is folded into quarters and placed on the desk. The stone is inserted by lifting the edge of the cloth and cleaned by rubbing it in the cloth between the thumb and forefinger. The textured side is used for diamonds and it should be kept meticulously clean. Leave the cloth folded over so the diamond side is never touched. Store it in a plastic baggie to protect it when not in use. Periodically, it must be cleaned, and this is done by hand washing it in dishwashing detergent and then rinsing thoroughly with water and allowing it to air-dry. Once the stone is clean pick it up in the tongs; do *not* touch it again with the fingers. Diamond has such an affinity for grease that even if the hands are freshly washed it will still pick up oils from the skin.

When examining any stone, keep a small brush handy to remove particles that remain on the surface.

Loupe

A loupe, or hand lens as it is alternatively known, should be standard 10X magnification and composed of a triplet lens corrected for spherical and chromatic aberration. This means that the three-part lens system has one lens responsible for magnification and two other lenses that correct for two separate types of distortion. Spherical distortion occurs when a lens does not bring all points in the same plane into focus at the same time. A lens corrected for this is termed *aplanatic*. Chromatic distortion occurs when a lens causes different wavelengths of light to be brought into focus at different distances from the lens, producing colour fringes around objects. A lens corrected for this is termed *achromatic*. When looking through the loupe at graph paper the lines should be in focus at the centre and at the ends without any widening of the lines and there should be no coloured fringes at the edges of the lines.

The primary purpose of the loupe is to examine internal and external gemmological features. The working distance of a loupe is 2.5cm, meaning the loupe must be held as close as possible to the eye and the stone held 2.5cm from the loupe. The loupe is held between the thumb and index finger, and in the opposite hand the tongs are held between the thumb and index finger with the handle resting inside the hand. It requires practice to develop a sturdy resting position and be able to maintain sharp focus. The best position is one in which the loupe hand touches the face and is steadied on the cheek, the stone hand touches the loupe hand, and both arms are locked on a solid surface such as a desk. Alternatively, with more practice, one can lock both arms at the sides of the body and reach in front of the face from the forearms; practise as much as possible until this becomes second nature. The light should be positioned just overhead, directed downwards and angled into the stone from the side such that it does not shine into the eyes. In this configuration inclusions appear bright against the dark background of the stone. The term for this type of lighting is dark-field. Learn to focus through the loupe without closing the other eye. The visual system is able to ignore input from the non-loupe eye and keeping it open makes constant loupe use more comfortable and cuts down on the detrimental effects that occur over time when extreme differences in visual demand are placed on each eye. When the loupe is not in use it should always be closed, or placed to rest on its side, not in the lens-down position. This ensures the lenses remain free of scratches and dust.

Stone tongs

Stone tongs, or tweezers as they are known alternatively, are available in a wide variety of styles. In reality it takes very little pressure to hold a gemstone in tongs provided that it has been picked up properly. Students often feel most comfortable in the beginning with locking tongs. Certainly this is advantageous in the classroom as the stone can be passed to others in the same position without fear of dropping it. With more experience, this can be done with any tongs. The most popular are those with slightly rounded tips that are grooved inside to help grip the stones. Tongs are available in black, brushed stainless and even lightweight titanium. Generally the lightest pair with not too 'stiff' an action enables the hand to use the tongs for many hours without developing cramps or muscle fatigue. Black tongs are not advisable for diamond grading as the black can reflect into the stones. Tongs should always be kept clean.

Picking up a stone is easiest when it is lying table-down on a firm surface. In any other position it may not be secure in the tongs. The tongs are held between the thumb and forefinger, parallel to the surface, and the tips are guided to either side of the stone at the girdle. Gently squeeze the tips to grip the stone. A light touch is all that is required. Do not apply excess pressure as this can cause the stone to pop out of the tongs. It is easier to pick up oval and rounded fancy-shape stones across the width, and squarish shapes corner to corner. Although the instinct is to pick up squarish shapes along the sides, in this orientation the sides of the stone do not parallel the tongs and they grip only the near or far end corners, so they are not held securely.

An alternative variety, known as corn tongs, has bent wires that emerge from the tip when the mechanism on the other end is pushed in. In this case the wires are positioned around the stone and the mechanism is released, allowing the wires to retract so the bent ends of the wires hold the stone. Corn tongs obstruct the view of the stone much more than proper stone tongs because they obscure four, or six, points around the girdle, and a student is ill advised to become reliant on them. It is best to learn to use stone tongs with confidence as the two points on the girdle obscured by the stone tongs can easily be checked by a minor repositioning. This is effortlessly achieved with practice.

Microscope

The microscope in its most essential form is simply a convenient carrier for a compound system of lenses, means for adjustment of their focus, and for control of the illumination of the specimen under examination. The primary function of the microscope is to give an enlarged image of an object set at its focus. The instrument performs, with significantly greater magnifying power, all the duties of the loupe; however, it is a much more versatile instrument with many additional applications. The price range of microscopes is considerable. The more complex the instrument the higher the cost, but an instrument suitable for gemmology may be rather basic.

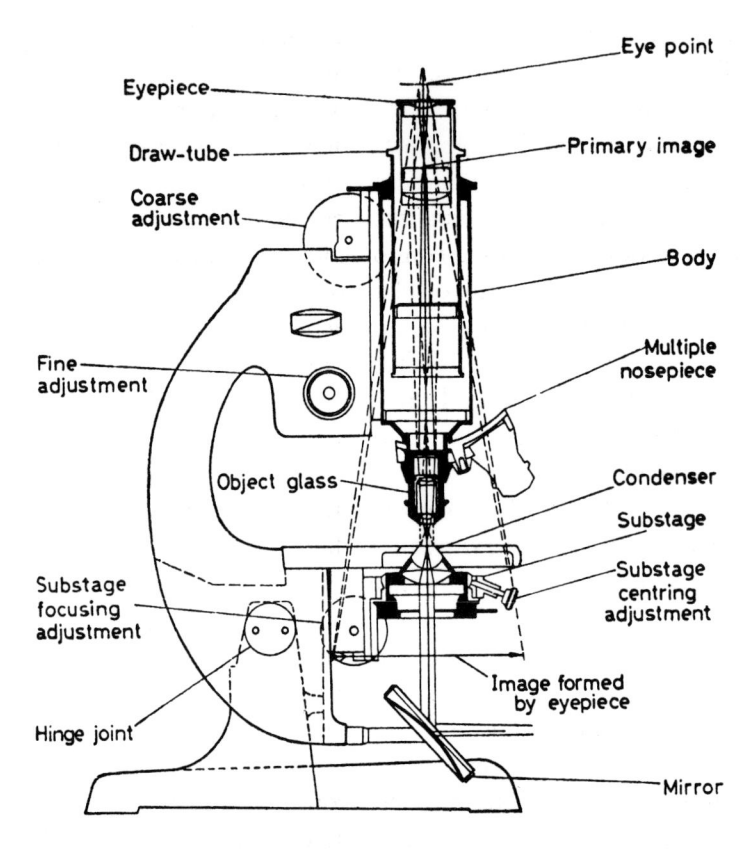

Figure 10.1 An illustration of one of the many models of microscope. Students are advised to learn the theory and uses of the microscope, but are not required to draw a diagram.

LOUPE AND MICROSCOPE

Microscopes have two lens systems: the lower group is the *objective* and the upper group is the *ocular*. Like the lenses of the loupe, these have been corrected for spherical and chromatic aberration. Gemmological microscopes are exclusively binocular. In a binocular microscope that does not have separate objective lenses for each eye, the image from the single objective lens system is split into two, allowing for two separate eyepieces that can be independently focused.

Magnification power is determined by the combined powers of the eyepiece and the objective lens. When higher magnification than this is required, auxiliary objective lenses are available which can increase the microscope's standard maximum magnification by 150–200%. A suitable maximum for gemmological microscopes is 60X to 80X, although most examinations are made at 15X to 45X and diamond grading is carried out at 10X. If the instrument is stereo-zoom the entire range of magnification can be accessed, not just pre-established increments. Binocular stereo microscopes that have two separate objective lens systems allow for true 3-D imaging. The greater the magnification, the smaller the field of view and working distance – that is, the distance between the objective lens and the specimen. An adequate working distance is about 5cm. This is comfortable for all general work, for the focal length is wide enough to clear the stone comfortably with the addition of a bright and wide field of view. Should the stone under test be set, for example in a ring, the shank of the ring prevents the objective from being lowered sufficiently to 'get into the stone'. In this case a longer focal length will be required to give clearance unless the microscope has an 'open well' dark-field.

The microscope stand is a heavy metal frame consisting of a *foot*, a limb on which the *body tube* is attached, and a platform called the *stage*. It is convenient to have a tilting base so the instrument may be used vertically or in an inclined position as this reduces muscle strain during periods of extended use. The platform on to which the specimen is placed or held above has a circular opening in the middle through which light is reflected from the adjustable mirror to reach the specimen. The circular opening in the stage is usually equipped with some kind of substage condenser controlling the quantity and area of the light passing to the specimen and there is always an opaque barrier blocking light from entering straight into the objective. Restricting light spillage from around the stone makes viewing features much easier. Both lens systems are mounted into the body tube of the instrument and may be raised and lowered for focus by a rack and pinion movement of two milled heads on either side of the tube. A stone holder allows the gemstone to be examined securely in any position, leaving the hands free to adjust focus and make notes. Detachable stone holders are best as they allow the stone to be picked up from table-down on the desk, without having to manipulate around the workings of the microscope base.

There are three types of illumination used with the microscope: (1) light-field – in which the dark central barrier is removed and light is transmitted through the specimen from below and into the objective; (2) dark-field – in which light is blocked from directly entering the objective and is transmitted into the sides of the specimen by means of a radial substage mirror. With dark-field illumination, the stone is viewed against a dark background, making internal features more visible; and (3) incident or top illumination – in which light is directed on to the stone from above. Most microscopes are equipped with overhead lighting, but independent desktop sources serve the same purpose. If diamonds are to be colour graded, this light should be a colour-corrected fluorescent lamp, specifically one that simulates north daylight in the northern hemisphere without fluorescence stimulating ultraviolet wavelengths.

The microscope should be placed on a firm table. The stone is held in tongs or the stone holder and is illuminated and examined. Looking from the side, lower the objective lens as close as possible to the stone and then, looking through the eyepiece, raise the objective to find the focus. This avoids damage to the objective lens by accidentally lowering it on to the stone while focusing downwards. Begin at the lowest magnification to find the focus and adjust as necessary. Keep the lenses clean at all times, avoid touching them, and keep the microscope covered when not in use to stop it collecting dust.

Table 10.1 Microscope uses

Observation of internal/external gemmological features
1 Doubling of back facets
2 Wear on facet edges
3 Scratches or wear on facets
4 Sharpness or roundedness of facet edges

5   Joining of facet angles
6   Indentation on moulded facets
7   Fractures
8   Cleavage
9   Bubbles
10 Inclusions and features: trigons, etch marks, striations, flux, etc.
11 Colour zoning
12 Curved striae
13 Differences in lustre either side of demarcation line in doublet
14 Overflow of glass on moulded stones
15 Colour layers along the girdle of soudé stones held edgewise
16 Quality of polish
17 Clarity grading of diamonds at 10X
18 Colour flash in fracture-filled diamonds
19 Concentration of colour along facet edges or in cracks

To determine refractive index

20 Real and apparent depth
21 Becke line
22 Other immersion methods

In combination with polarizing media

23 Observe pleochroism
24 Determine isotropy or anisotropy

Miscellaneous

25 Photomicrography and videomicroscopy
26 As a light source, such as to colour grade diamonds
27 Perform hardness tests under magnification to reduce damage
28 Check thermal reaction of plasticized or waxed stones without damage
29 Replace the eyepiece with the spectroscope. One can focus on one stone in a cluster to get a spectroscope analysis
30 Use the diaphragm as a light source to candle pearls
31 With endoscope to check natural/cultured pearls
32 Approximate gemstone proportions with an eyepiece graticule

The microscope is the most versatile of the gemmological instruments. Microscope uses include, to a greater degree, all the uses of the loupe for internal and external characteristics, features and markings of a stone. These are listed in Table 10.1. Higher magnification makes these features easier to see and this is of extra significance in certain cases. Especially with doubling of the back facets, the microscope makes it possible to see doubling even in stones with low birefringence, such as corundum 0.008, which cannot normally be seen with a loupe unless the specimen is unusually large. With stones in the RI range of 1.62–1.63, birefringence is a good distinguishing factor: tourmaline is 0.18, topaz is 0.008, and apatite is 0.002. This is more easily assessed with the microscope.

Immersion cells

When the light is adjusted so that the maximum amount of light enters the objective lens and one looks down the eyepiece, the field of view is seen to be quite bright. If a gemstone is viewed it will be seen that even when the microscope is carefully focused, it is difficult to see features inside the stone. This is due to the stone's polished facets reflecting back a significant amount of light, causing a distraction that makes it more difficult to see internal features. To a certain extent, adjusting the light can reduce these surface reflections; however, the best method is to view the stone when it is immersed in a glass vessel filled with high RI fluid. Ideally, this liquid should have nearly the same RI as the stone being examined and the two most commonly used are di-iodomethane and monobromonaphthalene. The closer the RI of the liquid approaches that of the stone, the closer the reflection of light from the stone's surface comes to being zero. Unless the stone is deeply coloured, it will become virtually invisible in a liquid of the same RI as itself.

Immersion cells are special fused glass cells. Clean the stone and the cell. Fill the cell and put the stone in with tongs. The stone must be fully immersed. Be certain to check that the glass of the cell is optically clean, meaning that it does not have its own swirls and bubbles that could be mistaken for features inside the stone. To avoid damage to the soft glass of the objective lens, look from the side and bring the objective close to the liquid and then focus upwards to avoid racking the tube down into the liquid. Slowly racking up will enable the observer to view the inside of the stone from bottom to top, and the stone can be 'worked through' without any danger of damaging the cell, the microscope or the stone. Dark-field lighting is best for viewing inclusions with immersion cells. Light-field lighting is better for curved striae and colour zoning. It helps in this case to place white paper below the immersion cell.

Remember that porous stones or those with surface-reaching feathers must not be immersed in high RI liquids. These liquids are expensive and hazardous in any case and it is sometimes just as suitable to use water, mineral oil or alcohol. Any liquid is helpful to some degree.

The inclusions seen in gemstones may be divided into three types: protogenetic, syngenetic and epigenetic. *Protogenetic* or pre-existing inclusions are those present before the formation of the host crystal, and such inclusions are usually seen to be solid crystals. *Syngenetic* or contemporary inclusions are those that co-crystallized with the stone and their chemistry reflects the environment of formation for both the stone and the inclusion. They may be drops of mother liquid or crystals that have grown at the same time. *Epigenetic* inclusions are those that have formed after crystallization of the stone. These include features such as cracks in which the mineral-rich liquid included in the cracks has crystallized out or, by alteration, an ex-solution inclusion occurs by chemical actions.

With the introduction of so many and varied synthetics and simulants it is no longer possible to assess a gemstone simply with the loupe at 10X. Most instances call for higher magnification. Learn the features of natural and synthetic stones, and the features of imitations and treated stones. Some inclusions are so typical of a gemstone that they may identify it. The 'horsetail' inclusion of demantoid garnet and the 'centipede' inclusion of moonstone are both diagnostic. In other cases they may indicate the country of origin such as the three-phase inclusions that are cavities containing a liquid, a bubble of gas and a crystal, which are evidence that the stone is an emerald from Colombia. The curved colour bands so indicative of synthetic corundums are not seen in the synthetic spinels, but in both cases the included bubbles are of typical types and may be round, tadpole-shaped or flask-shaped. Bubbles in glass are nearly always round or elliptical in shape, and glass often shows swirl markings. Details of these inclusions and features are given in the respective sections on gem species, synthetics, imitations and gemstone enhancements.

Additional uses of the microscope

Polarizing media can also be used in combination with the microscope to determine isotropy or anisotropy in the absence of a polariscope. The polars, either two Nicols or two polarizing filters, are used in conjunction with the optical train of the microscope. One polar, designated the polarizer, is inserted between the light source and the stone. The second polar, designated the analyser, is attached to the objective. They are placed in the crossed position and the substage condenser is usually removed from the microscope and parallel polarized light is used. The best results are found when the stone is immersed in an RI liquid such as monobromonaphthalene because this prevents surface reflections from the inclined facets or the exterior light. The rotating stage is then slowly turned through a complete circle. This will give similar results to those obtained with a polariscope: isotropic stones will remain dark for the complete rotation; anisotropic stones will give a four times light and four times dark display for the complete rotation, unless the direction of observation is parallel to an optic axis in which case it remains dark through the rotation. Polycrystalline materials appear light throughout the rotation. As with the polariscope, it is important to remember that isotropic substances under strain can show anomalous extinction (AEX) or 'tabby extinction' and this should not be confused with the sharp light/dark effect of true anisotropy.

In combination with polarizing media the microscope can be used to observe pleochroism. A polarizing Nicol or suitable filter is used and a coloured anisotropic stone is placed on the microscope stage. The vibration direction of the Nicol will 'select' the vibration directions of the stone so that as the stage is rotated, colour may vary every 90° due to differential selective absorption of each of the two rays in the stone. Thus, pleochroism may be seen in the absence of a dichroscope, but in this case the two colours are not seen side by side. Weak dichroism may be missed because the eye cannot make such delicate distinctions in the absence of a direct comparison.

Microscopes with a fine focusing adjustment graduated to read in hundredths of a millimetre are ideal to determine refractive index by the real and apparent depth method. This method, as well as others, is discussed in detail in Lesson 7.

Microscopes can also be used for photomicrography. Pictures taken in combination with the microscope allow a permanent record of a particular feature of a stone. Single-lens reflex cameras are popular. The lens is removed and it is mounted over the eyepiece and this allows the reflex image to be viewed and focused in the usual way.

A simple microscope with a substage condenser proves a very useful light source for the spectroscope. The stone is placed on the stage on a glass slide, the eyepiece is removed, and the low power objective used. The mirror, condenser and position of the stone should be adjusted to allow the greatest possible amount of illumination in the centre of the field. The focus must be raised so the field of view is an out of focus glare of transmitted light. The spectroscope can be rested on the body tube where the eyepiece normally is and a spectrum can be seen which is free of streaks and is well and properly illuminated. In this manner it is possible to get a spectrum from just one stone in a cluster ring. Additional microscope uses are presented in Table 10.1.

LESSON 11

Primary Gem Species

This lesson discusses the major gem species of the professional gemmology curriculum. It is important to study these gems by reading as much material as possible from a variety of sources, including journal articles. One must know all their properties as well as identification factors that separate these stones from one another, simulants and synthetics as well as detection of possible treatments. Some of this information has been given in other lessons. It is advised to make thorough notes and tabulate data for efficient reference and comparison.

Diamond

Diamond crystallizes from carbon at conditions of high pressures, greater than 45 kilobars, and high temperatures, greater than 1200°C. Most diamonds form in the upper mantle at about ~ 140–<400km in two distinct types of rocks: peridotite and eclogite. At the surface, diamonds are found in two kinds of deposits: (1) primary – typically funnel-shaped volcanic bodies, most often kimberlites, but more rarely lamproites and lamprophyres. These rocks have transported the diamonds upwards in magmas that erupt at the surface; and (2) secondary – deposits formed by the weathering and erosion of primary deposits, typically by river, stream, marine and glacier activity that displaces and re-deposits them as alluvial deposits in streambeds and terraces and as marine and offshore deposits in coastal areas.

Diamond and/or diamond-bearing rocks have been found on all seven continents. Diamonds have been known in Borneo and Indonesia since antiquity and their presence documented as early as the fourteenth century. India is another historical source, as is Brazil where diamond was found in 1725. Presently, the principal commercial deposits are in Africa – most importantly, Botswana, South Africa and Namibia; Russia; Canada; and Australia.

In the diamond structure, each carbon atom covalently bonds with four other carbon atoms producing equidistant tetrahedral bonding in which all bond angles are 109.5° and all bond distances are 0.1544nm. Diamond crystallizes in the isometric system – which is also termed 'cubic system'. The most common primary growth form is the octahedron, which often shows characteristic equilateral triangular surface features known as trigons. Cubes are also common, but rare in gem-quality stones. Combination forms such as the octahedron-dodecahedron are known. Diamond also occurs as macles, flattish triangular shapes produced by twinning. Many diamond crystals are frosted, coated or have a skin referred to as nyf.

Chemically, diamond is pure carbon although impurities have important influences on its properties. Gem diamond is divided into Types. Type I diamond has nitrogen as an impurity: Type Ia in groups and clusters and Type Ib as single atoms. Type II lacks nitrogen: Type IIa without impurities, and Type IIb with boron. Boron produces the blue colour that the Hope Diamond is famous for; it also produces electrical semi-conductivity. Type Ia, comprising 99.8% of all diamonds, constitutes the Cape Series with a characteristic absorption spectrum: a signature line at 415nm and other bands at 423nm, 435nm, 452nm, 465nm and 478nm. Type Ib shows a general strong rise in absorption in the blue. Type IIa has absorption in the yellow-green centred on 563nm and Type IIb exhibits no spectrum. Diamond exhibits photoluminescence: the response is sometimes patchy, but in general the LW response is stronger than the SW response. Various colours and intensities result from different body colour, Type and impurities. It is diagnostic in only one instance: blue fluorescence with yellow phosphorescence.

Diamond is isotropic and never exhibits birefringence, yet it almost always shows anomalous extinction (AEX) between crossed polars due to strain. Its refractive index is 2.417. Dispersion is 0.044, specific gravity is 3.52, the lustre is adamantine, fracture is conchoidal, and it exhibits perfect cleavage in the octahedral direction. In reality, fracture typically occurs in combination with cleavage, leaving a stepped surface that is somewhat splintery. Diamond is the most transparent crystalline material known, able to transmit a great deal of the electromagnetic spectrum, including X-rays.

The extreme hardness of diamond, 10 on the Mohs' scale, is its most famous property. It is a directional property in which variations can exceed 100:1. Graining, colour differences, and the presence and quantity of nitrogen also affect hardness. It is often said that diamonds from certain locations are harder than others; those of Invernell in New South Wales are reported to be the hardest, with India, Borneo and Brazil next hardest. Geographical differences are attributable to the presence of knots that slow the cutting and polishing process, thereby giving the impression the diamonds are harder. It is due to directional hardness that diamond can be cut at all. Diamond can only be cut by diamond, but never in its hardest direction. Diamond exhibits significant toughness, the ability to withstand disruptive stress; however, it is only extremely resistant to stress applied as a slow force, such as squeezing in a vice, and less able to stand up to sharp stress, such as force applied by a blow. This is often enough to induce cleavage and break the diamond.

Diamond is primarily thought of as a colourless gem and pure undeformed diamond is always colourless; however, colour can be caused by impurities, structural deformation and inclusions. As a result, diamond is naturally found in all colour varieties: red, orange, yellow, green, blue, violet, black, white, grey, purple, orange, pink and olive. These stones are referred to as *fancies* in the trade. All fancies are very rare in fine quality, some more than others: red, green and blue are the rarest. Fancy white diamond is actually a translucent milky-white colour, not to be confused with the common usage of white to refer to colourless material. Black diamonds are coloured by graphite inclusions, white by scattering, yellow by nitrogen, blue by boron, green by natural irradiation, and pink by plastic deformation. The finest stones of the red, green, blue and pink varieties can command over $1 million per carat.

Colour treatment of diamond is undertaken to enrich colour or to disguise undesirable tints. Surface treatments to mask yellow tinges and make the diamond appear whiter have been known for centuries. Various dyes, inks, stains, varnishes, enamels and foil backing have all been used. Irradiation and annealing produces additive colour that imitates fancy diamonds. HPHT treatment can produce and modify colour in the additive, or it can lighten body colour. Fracture filling reduces the visibility of open inclusions and laser drilling can improve the look of diamonds with dark inclusions.

Diamonds are graded for commercial purposes on the basis of the modern-day four Cs: carat, colour, clarity and cut. Carat refers to metric carat weight. Colour refers to grades of body colour ranging from the ideal colourless state (D–Z). Clarity grades describe the increasing presence of inclusions from the ideal flawless state (F–I_3). And cut grading assesses the effect of proportions and faceting quality on overall beauty and brilliance.

With experience it is possible to identify diamond on sight alone, as its characteristic lustre, brilliance and fire are not seen with any other stone. With the loupe, indications of hardness, quality of cut, confirmation of isotropic nature, and inclusions such as diamond in diamond, graining and naturals may be evident. Diamond is almost never identified by its RI or SG. The read-through test and the tilt test are useful, as is visual optics. A number of meters and testers have been marketed that may differentiate diamond from its imitators on the basis of select physical properties, typically reflectivity/reflectance, thermal conductivity and electrical conductivity.

In 1954 General Electric succeeded in synthesizing diamond in the laboratory. Initially these stones were small, expensive to produce and not of gemmological concern; however, in recent years the price of gem-quality coloured synthetic diamonds has become competitive and a number of companies are producing for the marketplace. To date, colourless synthetics are not competitive although research and development is ongoing.

In addition to gem diamond, 'boart' or 'bort' is the term used for a massive aggregate of tiny diamond crystals, usually dark in colour, used as an industrial abrasive. Carbonado is a massive form of impure diamond, a natural aggregate of cryptocrystalline diamond, graphite and amorphous carbon. It is also used industrially, most notably in diamond drills.

Corundum

This is oxide of aluminium (alumina), with the chemical formula Al_2O_3. The varieties are ruby (red) and sapphire (blue). The other colours are yellow, green, orange, pink, purple, violet, black, brown, grey and colourless. These are termed sapphire, with the colour as a prefix. Star-stones show 6-rayed, occasionally 12-rayed, stars. Asterism is common.

Chatoyancy is known but rare. Colour-change stones are rare. Fancy sapphire is the trade name applied to all colours of corundum except red and blue. Pure corundum is colourless; colour is due to impurities. Ruby is coloured by chromium, but the colour is modified by ferric iron; sapphire is coloured by iron and titanium; yellow sapphire by iron; and green sapphire by iron or iron and titanium. Emery is an impure form of corundum used as an abrasive.

Corundum is trigonal and habit varies with species and location: ruby typically as tabular hexagonal prisms terminated with basal plane; sapphire typically as hexagonal bipyramids with lateral striations forming a barrel-shape. Hardness is 9, but sapphires are slightly harder than rubies. The lustre is vitreous and the RI is 1.76–1.77 with a birefringence of 0.008 to 0.010. Uniaxial negative. SG is 4.00 and dispersion is 0.018. No cleavage, but there is false cleavage known as parting. Fracture is usually uneven, but may be conchoidal. Dichroism is strong in most sapphires: violet-blue and greenish-blue; and in fine coloured rubies: orangy-red and deep purplish-red. In both cases the best colour is in the direction of the optic axis. Stones are cut in various styles including: step, mixed, cabochon, briolette and sometimes beads.

Chromium-coloured stones, including ruby, pink and violet sapphire, luminesce strong red under UV; this diminishes in proportion to iron content that quenches the response. Almost all blue sapphires are inert, but Sri Lankan show red/orange due to a trace of chromium. Yellow and green are typically inert. Rubies are typically fluorescent red under the Chelsea filter. Both ruby and sapphire have diagnostic spectra. The chromium absorption spectrum of ruby is a strong doublet in the deep red, usually seen as a single bright fluorescent line at 693nm, with fainter lines in the orange-red at 668nm and 659nm, strong general absorption of yellow and green from 500nm to 610nm, and three lines in the blue at 476.5nm, 474nm and 468.5nm with a strong cut-off in the violet. Blue and green sapphires, due to ferric iron, show a series of three bands in the blue, the strongest is at 450nm and others at 460nm and 471nm. With ruby, various heat treatments can improve or create asterism, remove purplish or brownish coloration, and diffuse colour on to the surface in combination with colouring agents. With sapphires, various heat treatments can reduce rutile clouding, produce yellow to orangy-yellow colours, lighten very dark blue and green colours, and diffuse colour on to the surface in combination with colouring agents. Most sapphires are heat-treated to lighten or darken the colour. Oiling, dyeing and glass infills are used for ruby and sapphire.

The inclusions of ruby and sapphire are some of the best means of identification and separation from synthetics. A delicately woven network of microscopic hollow tubes and needle-like crystals of rutile crossing at 120°, known as silk, is diagnostic for natural and characterizes the finest stones: Myanmar ruby and Kashmir sapphire. In ruby, swirls of deep colour, described as 'treacle', and zircons with stress haloes are also diagnostic and characteristic of Myanmar stones. Thai rubies lack silk, but often show crystal inclusions such as almandine garnet, yellow apatite or pyrrhotite crystals surrounded by liquid feathers. Other crystal inclusions found in ruby include spinel, zoisite, magnetite, pyrite and flakes of biotite mica. In sapphire, Kashmir stones may contain green tourmaline or green mica. Myanmar, Thai and Australian stones may contain feldspars, calcite, apatite and zircon. Small red inclusions of uranium may be found in rubies from Cambodia and graphite may be found in Tanzanian stones. Sapphires characteristically show pronounced colour zoning.

Corundum occurs in gravels and clays derived from an impure limestone, recrystallization being caused by metamorphic influences of heat and pressure. Gems are usually recovered from alluvial gravels. The finest rubies are found in Burma, now Myanmar, from the Mogok district. Rubies from Thailand are more purplish, Sri Lankan stones are lighter in colour but a purer red, and this can merge into the variety known as pink sapphire. Greenland produces ruby from igneous rocks of a fine red colour. Rubies are also found in Australia, Pakistan, Afghanistan, Zambia, Cambodia and Kenya. The finest sapphires, an intense medium, slightly violet, blue, are from Kashmir in the north of India. Stones from Thailand vary from finest blue to dark blackish, Sri Lanka produces greenish- to violetish-blue, which may be lighter in colour as well as many fancy colours. Australia is the source of the best fancy greens and also produces blues, which typically have greenish overtones ranging to a dark inky blue. Those from Montana, in the United States, have a steely blue colour. Sapphires are also found in Myanmar in association with ruby, and in Cambodia and Zimbabwe.

Careful RI work obtaining proper birefringence typically identifies corundums from other minerals. Spectroscope is diagnostic in ruby and blue and green sapphire. Always check for composite stones and treatments such as diffusions, oils and dyes, and glass infilling.

Beryl

This is silicate of aluminium and beryllium, with the chemical formula $Be_3Al_2(SiO_3)_6$. There is isomorphous replacement of the beryllium by alkali metal oxides and of the aluminium by chromic or ferric oxides. The principal commercial varieties are emerald (pure green) and aquamarine (pale blue to sky blue to bluish-green). Other colours may be referred to by their species name or by their colour: rich yellow or golden colour (heliodor); pink (morganite); red (bixbite); green other than emerald (green beryl); and colourless (goshenite). Star stones are known. Maxixe, a variety of blue beryl from the Maxixe mine in Brazil, has a medium to dark-blue unstable colour that fades slowly and permanently in sunlight and strong light. Some aquamarines and yellow beryls show chatoyancy. Asterism is rare.

Beryl crystallizes as six-sided prisms in the hexagonal system. Emerald is usually terminated with the flat faces of the basal pinacoid. In aquamarine and other varieties small pyramidal faces bevel the junction of the basal face and the prism faces. Aquamarine crystals can grow very large and are often striated parallel to the prism edge. This can be so pronounced that it obscures the hexagonal outline and the crystal appears as a ribbed cylinder.

For beryl in general, hardness is 7.5: emerald is about 7.5 or slightly less and the others are a little harder. Cleavage is weak and indistinct parallel to the basal plane and fracture is conchoidal. SG averages 2.70 and the full range for the species is 2.67 to 2.90 with emerald ranging from 2.67 to 2.80, and aquamarine ranging from 2.66 to 2.80. The lustre is vitreous and RI is 1.57–1.58 with a birefringence of 0.006. RI and birefringence vary slightly by species and the constants of emerald vary by location. Uniaxial negative. Dispersion is 0.014.

Emerald is coloured by chromium replacing aluminium ions in the crystal lattice. There is often some vanadium present as well as some iron, and these modify the colour. Emerald is brittle and gem-quality material usually has inclusions visible with the naked eye. The finest emeralds are Colombian, where it occurs in calcite veins, from famous mines such as Chivor and Muzo. In Brazil and the Ural Mountains of Russia it occurs in mica schists. Brazilian emeralds are smaller and of lower quality than the finest Colombian stones; some are very dark green with profuse inclusions and some are coloured by vanadium and have little, if any, chromium. Emerald inclusions are very characteristic of their deposit: Colombian – three-phase inclusions with a liquid, vapour bubbles and one or more crystallites; groups of pyrite crystals; and wing-shaped healing fissures; Russia – actinolite needles; India – 'twinned' two-phase inclusions; Zambia – black clouds of magnetite crystals; orthorhombic prisms of chrysoberyl; Australia – ilmenite; Brazil – growth tubes parallel to c axis; Zimbabwe – fine tremolite fibres; and South Africa – shiny silver molybdenite.

Emerald shows a similar chromium spectrum to ruby characterized by sharp lines in red. The most prominent bands are the doublet of 683nm and 680nm and it is more widely spaced than in the ruby, lines at 646nm and 662nm, weak diffuse central absorption between 580nm and 625nm, weak lines in the blue, and absorption of the violet. Under the Chelsea filter most emeralds show pink to red. This is a fluorescent red light and may be diminished by the presence of iron. Emerald shows strong dichroism: blue-green and yellow-green.

The finest aquamarine is sky blue and, even if deep, never competes with the hue of fine sapphire. Aquamarine is coloured by iron and shows strong greenish-blue through the Chelsea filter. The absorption spectrum is not very pronounced with indistinct lines at 456nm and 537nm and a strong line at 527nm. It is inert to UV. Aquamarine can be found in large flawless crystals although smaller and paler stones are much more common. Practically all fine blue aquamarine is heat-treated greenish-yellow or brownish-yellow material; the change is permanent and acceptable. Aquamarine as well as beryls other than emerald occurs mostly in coarse-grained rocks known as granitic pegmatites. Aquamarines are found in Brazil, Russia, California, Madagascar and Myanmar. Aquamarine is typically eye-clean but under the microscope inclusions may be seen: thin rod-like 'rain', acicular inclusions which can give chatoyancy, and 'chrysanthemum' inclusions appearing as flat snowflake-like inclusions. The dichroism is strong in aquamarine if the colour is strong: deep blue and colourless.

Morganite, pink beryl, is coloured by manganese. It has higher constants due to rare alkali earth metals. Impurities in the stone tend to raise the SG to average 2.80 to 2.90 and the RI to 1.58–1.59 with a birefringence of 0.008 or 0.009. Dichroism is distinct: pale pink and deeper bluish-pink. It exhibits no spectrum, shows weak lilac under UV, and intense but not bright crimson under X-rays. It is found in pegmatites in California, Brazil and Madagascar. Bixbite, red beryl, is also coloured by manganese. It is found in Utah but only in small sizes, and the finest gem-quality stones are 2ct and under. Dichroism may be strong: red and bluish-red. Green beryl and heliodor are coloured by iron. Both are inert to UV and found with aquamarine in Madagascar, Brazil and Namibia. Heliodor shows weak dichroism: golden-yellow and stronger yellow. Green beryl is typically weak: green and lighter green. Goshenite is found in Massachusetts. Inclusions in varieties other than emerald include: liquid, two-phase or tabular inclusions, and feathers of negative crystal cavities. Hollow straight or liquid-filled tubes are also common. Most beryls are step cut, some as cabochons and fancy shapes.

RI and SG may identify all beryls as no other comparable stones have similar constants. Always watch for composite stones, including soudé type, as well as glass. Synthetic emerald is produced by a variety of means, discussed in Lesson 12, and can be differentiated by inclusions and often by RI.

It is safe to assume that all emeralds are oiled to conceal cracks that reach the surface, which improves the overall transparency of the stone. Unless green colouring or a bonding agent is added, the practice is generally accepted; however, oiling is unstable and will dry out and discolour over time. And ultrasonic cleaning and solvents will cause the oil to deteriorate and the flaws to reappear. Ultrasonic cleaners work by vibrations that generate tiny bubbles that collapse back into the liquid to 'scrub' what is being cleaned. These cleaners are dangerous for emerald because expanding and collapsing bubbles inside inclusions of the emerald can make the inclusions larger or break the stone apart. Heat treatment improves the colour of morganite by removing yellow tints. This is common, stable up to 400°C and acceptable. Irradiation produces yellow beryl from colourless. It is undetectable, stable to 250°C, and acceptable. Irradiation produces dark blue Maxixe type from some colourless to light pink beryl; it is unstable, fades in light and heat, and can be detected by spectrum.

Topaz

Topaz is aluminium silicate with some hydroxyl and/or fluorine, with the chemical formula $Al_2(F,OH)_2SiO_4$. The (F,OH) indicates that the hydroxyl and fluorine can substitute for each other at the same site in the crystal lattice due to isomorphous replacement. Although pure end members of the isomorphous series are theoretically possible, there is always some fluorine and some hydroxyl present. In the past, the name topaz was used indiscriminately for many yellow stones, in particular yellow citrine from Brazil. This caused true topaz to be referred to as precious topaz. The name topaz should only be used to refer to the mineral topaz. Similarly, the stone smoky topaz is a misnomer too; it is smoky quartz and should always be referred to as such.

Topaz crystallizes in the orthorhombic system. Crystals are commonly prismatic in habit, with prism faces vertically striated; one end is usually terminated with pyramidal or dome-shaped faces and in some cases two of the dome faces can be so enlarged they meet at the top of the crystal to form a ridge that gives a chisel-shaped appearance to the crystal; the other end is typically not terminated but consists of the basal cleavage plane formed when the crystal breaks – as a result of easily developed perfect basal cleavage at right angles to the length of the crystal, from the rock on which it had grown. In cross-section, crystals are lozenge or diamond shaped. If the second-order prism is predominant, the outline is pseudo-tetragonal or square.

Colourless and shades of pale blue are most common; yellow through to brownish-yellow/yellowish-brown to orange (often described as sherry brown), and blue to green also occur naturally. Strong blues are most likely due to irradiation. Red and pink colours are rarely found naturally. Most pink topaz is heat-treated brown material; the red is likely natural; and salmon pink is a result of overheating. The most prized topaz is the vivid pinkish-orange to red-orange variety known as Imperial topaz. Pink and red topaz are coloured by chromic oxide as are Brazilian browns; other colours, including blues, greens and other browns, are due to colour centres. Orange is due to colour centres and chromium. Trichroism is distinct in well-coloured stones, especially in pink topaz: colourless and two shades of pink; sherry brown: honey yellow, straw yellow and pinkish-yellow; and blue: colourless, pale blue and blue.

Due to isomorphous replacement involving the (F) and (OH), the constants are somewhat variable, but normally gem-quality topazes fall into two groups: (1) fluorine-rich – typically colourless and blue, but also some yellows and browns: SG 3.56, RI 1.61–1.62, and birefringence 0.010; and (2) hydroxyl-rich yellow, brown and pinks: SG 3.53, RI 1.63–1.64, and birefringence 0.008. Fluorine raises SG but lowers RI, whereas hydroxyl raises RI but lowers SG. Topaz is biaxial positive. Dispersion is 0.014. Hardness is 8, it is said to feel slippery, and lustre is very bright vitreous. Topaz has perfect cleavage parallel to the basal plane. Fracture is conchoidal. The only topaz that may show an absorption spectrum is pink topaz, which may exhibit a weak chromium line in the red at 682nm. Blue topaz shows greenish blue under the Chelsea filter, and this is much weaker than the strong blue of aquamarine. Luminescence varies with type: fluorine-rich exhibit weak yellow or green-yellow under LW, very weak similar colours under SW, and greenish-white to violet-blue under X-rays; hydroxyl-rich exhibit strong orange-yellow to LW, very weak similar colours under SW, and brownish-yellow to orange under X-rays.

Topaz occurs in cavities in highly acid igneous rocks such as granite and also in zones of contact metamorphism and in granitic pegmatite dykes. Gem topaz is predominantly from pegmatites in association with minerals such as apatite, beryl, mica, quartz and tourmaline. Topaz is found as water-worn pebbles in river gravels. The principal localities for topaz are: Brazil, notably Ouro Preto, which is a source of Imperial topaz, as well as sherry brown, fine yellow, blue and pink; African countries such as Namibia, Zimbabwe, Tanzania, and Nigeria are noted for blue and colourless used for irradiation to blue. Other sources are: Australia, Russia, Sri Lanka, Myanmar, Mexico and Scotland.

A great deal of topaz is inclusion-free. Tiny cavities containing two or three immiscible liquids are common in white and blue fluorine-rich topaz; they are often drop shaped with bubbles one inside the other, and they may also include a solid phase. Cubic crystals, possibly of fluorite, have been seen in Nigerian stones. Hydroxyl-rich stones usually have long tube-like cavities along the length of the c axis. Topaz from pegmatites can show crystalline inclusions including spessartine garnet and quartz.

Its RI, birefringence and optic character and sign can identify topaz. Yellow tourmaline is close but proper birefringence easily separates it and danburite is close in birefringence but is optically negative. With experience it is possible to notice the lustre is higher on topaz than quartz. Glass may have an RI close to topaz, but it is non-birefringent. Topaz is less likely to be patchy or zoned in colour than is citrine. Tourmaline, aquamarine, spodumene, danburite and quartz, those stones most likely to be confused with topaz, all float on di-iodomethane, SG 3.32, and topaz sinks.

Irradiation produces: deep blue topaz from lighter blue or colourless material. The stones must be heated to remove unwanted yellow or brown colour components. This is generally stable, undetectable and an accepted practice. London Blue refers to medium to dark greyish-blue, which is often described as somewhat inky. Swiss Blue refers to darker blues without inkiness. Sky Blue refers to the brighter lighter blues. Heat treatment produces pink from brown material.

Topaz is piezoelectric and pyroelectric, meaning it becomes electrically charged when subjected to mechanical stress such as friction, or heat. Topaz is typically cut in the mixed cut style as pears or ovals to maximize yield from the prismatic crystals. Other fancy cutting styles, including laser-cut fancies, are common in blue topaz as are briolettes. Topaz is made synthetically but not for commercial jewellery. Any so-called synthetic topaz is synthetic corundum or synthetic spinel of suitable colour.

Spinel

This is oxide of magnesium and aluminium, with the chemical formula $MgAl_2O_4$. Spinel is part of an isomorphous series in which the magnesium may be replaced by ferrous iron or manganese and the aluminium may be replaced by ferric iron or chromium. Spinel is the only member of the series that is gemmologically significant. Spinel crystallizes in the isometric system, commonly as octahedrons, sometimes modified by dodecahedral faces. Spinel twins on the octahedral plane and this type of twinning, which forms the macle, is known as 'spinel twinning'. Spinel is commonly found as water-worn pebbles. Commercially important are: red – the best is a deep intense red like the colour of fine ruby, but more common are paler reds and those that are orangish, brownish or purplish; and blue – varies from rich blue like that of fine sapphire to paler greyish-blue or greenish-blue. Fine cobalt blue spinel does occur, but is very rare. Red and pink are coloured by chromium and blue by ferrous iron. Additionally, violet to purplish stones are common; bright orange; green colours – true green is rare; most are very dark green due to excess iron and these are known as ceylonite or alternatively pleonaste, and are essentially black; colourless stones are never truly colourless – there is always a trace of pink. Star spinels with four or six rays are rare.

Hardness is 8. SG averages 3.60; pink stones have lower values and iron-rich stones such as ceylonite average 3.80. RI is 1.718, usually expressed as 1.72, but chromium-rich reds may reach 1.74, zinc-rich varieties such as pale to dark blue gahnospinel may reach as high as 1.753, and ceylonite may be 1.77 to 1.80. Spinel is isotropic and therefore shows no birefringence or pleochroism. Lustre is vitreous. Dispersion is 0.020. There is no cleavage, fracture is conchoidal, and the stone is brittle. Spinel is cut in step, mixed and fancy shapes.

Red spinel has a diagnostic absorption spectrum: a characteristic group composed of four closely spaced lines in the red centred on 680nm. The two brightest of these fluorescent emission lines, caused by chromium, are at 686nm and 675nm, usually accompanied by up to eight other lines producing what is described as the *organ pipe* spectrum: 632nm, 642nm, 650nm, 656nm, 665nm and 684nm. There is broad absorption of the green centred on 540nm and absorption of the violet below 450. Blue spinel coloured by iron usually shows a moderately strong band in the orange at 630nm, a stronger and narrower band at 480nm, and strong broad band in the blue at 459nm. Also, fainter bands may be seen at 433nm, 443nm, 508nm, 555nm and 585nm due to ferrous iron. The very rare natural cobalt-coloured blue spinel shows the typical cobalt spectrum with three bands: one in the green at 540nm, one in the yellow at 580nm, and one in the orange-red at 635nm and weak absorption from 430nm, and usually some iron bands are faintly noticeable.

Natural red and pink spinel fluoresce red to LWUV and the response is much stronger than under SWUV. It is moderate red under X-rays and there is no phosphorescence in any case. Dark-blue spinel with iron is inert to UV and X-rays. Paler blue or violet spinel is green under LW and X-rays and inert under SW.

Spinel typically forms at high temperatures from metamorphosed impure limestone. It has also been found in igneous rocks. Spinel is generally found in association with corundum. Myanmar is a source of fine red, pink and blue. Sri Lanka is known for blues and violets but produces a full range of colours. Other sources include: Kenya, Tanzania, Australia, Brazil, Afghanistan, and Pakistan. Thailand, contrary to many reports, is only a source of black spinel.

Spinel is characteristically a clean stone with few inclusions but, when seen, included crystals with surrounding iridescent stress fractures, known as 'spangles', are common, as are minute octahedral crystals either as individuals or in rows forming what appears as a fingerprint. Thin films containing iron staining and zircon haloes are also found. Liquid inclusions are rare.

Synthetic spinel is common, but the colour does not match the natural varieties. The lighter blue simulates aquamarine and can be detected by showing red through the Chelsea filter. Synthetic red spinel is rare and lacks the organ-pipe spectrum of the natural. It is not often produced because the correct ratio of magnesium and aluminium must be obtained for the chromium to yield the red colour. If trace chromium is present in synthetic spinel with the typical excess of aluminium, then the resulting colour is an unattractive greyish-green, rather than the desired red, presumably because of lattice distortion by the excess aluminium. Synthetic spinels, except for the red, have a higher RI of 1.728 and show tabby extinction between crossed polars. It is discussed in detail in Lesson 12.

Garnet

Garnet is an isomorphous group of gemstones, with the general chemical formula $A_3B_2(SiO_4)_3$. All are double silicates in which A is divalent calcium, iron, magnesium or manganese and B is trivalent aluminium, iron or chromium. Garnets are a classic example of isomorphous replacement where metal ions having similar ionic radii and chemical affinity can interchange with one another to produce an isomorphous series. Garnets are divided into two series: (1) *pyralspite series* – pyrope, almandine and spessartite; and (2) *ugrandite series* – uvarovite, grossular and andradite. Almost all garnets are mixtures as there is reasonably complete isomorphous replacement within a series, but little or only partial isomorphous replacement between series.

Garnets crystallize in the isometric system, often as euhedral crystals. Rhombic dodecahedra and trapezohedra are common as are combinations of the two. Water-worn pebbles may appear almost spherical. The lustre of all garnets is vitreous to resinous except demantoid, which is said to be sub-adamantine. There is no cleavage but sometimes distinct parting parallel to the dodecahedral crystal face. Fracture is conchoidal to uneven. Garnets are inert to UV. All are isotropic and therefore non-birefringent and non-pleochroic; the exception is hydrogrossular, which is polycrystalline. Typically garnets are red under the Chelsea filter.

Pyrope is magnesium aluminium silicate – $Mg_3Al_2(SiO_4)_3$. It is almost always mixed with some almandine. Pure pyrope is colourless. The usual gem colour is blood red, but it may be medium to dark red and tinted with yellow or purple. Hardness is 7.25, SG is 3.65 to 3.78, RI is 1.73 to 1.76, and dispersion is 0.022. Pyrope exhibits moderate magnetism. Spectrum may show almandine or spessartite bands: a cut-off at 440–445nm, broad band at 570nm, a pair of lines centred on 685nm, and weak band at 505nm. It is found in mica schist, and also known in South Africa and Botswana in association with diamond. Also found in Russia, Sri Lanka, Tanzania and Zimbabwe. Characteristic inclusions are rutile needles, apatite crystals, zircons with tension haloes, and rounded low relief crystals.

Almandine is iron aluminium silicate – $Fe_3Al_2(SiO_4)_3$ – with variable amounts of magnesium, calcium and manganese. Colours are dark intense red to purplish-red. Some show asterism with four- and six-rayed stars. Hardness is 7.5, SG is 3.95 to 4.20, RI is 1.78 to 1.81, and dispersion is 0.027. Almandine exhibits a characteristic spectrum of three strong bands: 505nm, 526nm and 576nm with weaker bands at 462nm and 617nm. Values of SG and RI between pyrope and almandine, SG 3.78 to 3.95 and RI 1.76 to 1.78, are called intermediate garnets or pyrope/almandine series. The garnet known as *rhodolite* is two-thirds pyrope and one-third almandine. It is a light to dark purplish-red or reddish-purple. It must have some purple component. Hardness is 7.25, SG is 3.83 to 3.90, and RI is 1.75 to 1.78. It may exhibit a moderate almandine spectrum and coarse needle-like inclusions. Rhodolite was originally found in North Carolina and sources include Kenya, India, Sri Lanka and Madagascar. Almandine is found in mica schists in India, Sri Lanka, Brazil, and North and South America. Characteristic inclusions are the same as for pyrope. Almandine exhibits strong magnetism.

Spessartite is magnesium aluminium silicate – $Mn_3Al_2(SiO_4)_3$ – often with some isomorphous replacement by almandine. Hardness is 7.25, SG is 3.90 to 4.20, RI is 1.79 to 1.81, and dispersion is 0.027. Colour is reddish-orange to yellow-orange-brown. Spectrum shows manganese lines in the blue at 412nm, 424nm, 432nm, 462nm, 485nm and 495nm, which may merge to a cut-off at 435nm and almandine iron bands at 505nm, 526nm and 576nm. This unusual spectrum is very useful for identification. Fingerprint inclusions are characteristic and a good separation from other orange stones. Also seen are wavy feather-like liquid inclusions and negative crystals. It is usually found in granitic pegmatites in Bavaria, Sri Lanka, Brazil, Madagascar, Australia and California.

Additional to the pyralspite series is the orange-coloured malaia garnet. It is a pyrope/spessartite garnet with SG 3.88 and RI 1.765. The spectrum, caused by vanadium, shows three bands: 455nm, 495nm and 580nm. Rutile, pyrite and apatite crystals are common inclusions. There is also a greenish-blue to magenta-red colour change pyrope/spessartite garnet from the Umba River in Eastern Africa. A concentration of actinolite fibres partly dissolved into dots is a characteristic inclusion.

Uvarovite is calcium chromium silicate – $Ca_3Cr_2(SiO_4)_3$. It is a rich green colour, due to chromium, but has not been found in crystals large enough to cut and is not commercial. Hardness is 7.5, SG is 3.77, RI is 1.87 and dispersion is 0.022. It may exhibit a chromium spectrum. It is found with chromite in California and in granular limestone in the Ural Mountains.

Grossular is calcium aluminium silicate – $Ca_3Al_2(SiO_4)_3$. Pure grossular is colourless. Massive grossular garnet is opaque but the gem varieties *hessonite* and *tsavolite* (or *tsavorite*) are transparent. Hessonite is orange-red to honey yellow due to iron and manganese. Hardness is 7.25, SG is 3.65, RI averages 1.74, and dispersion is 0.027. Hessonite exhibits moderate magnetism. Reddish-orange hessonite may show bands at 407nm and 430nm. Characteristic inclusions include myriads of apatite and calcite crystals and peculiar treacly streaks which give an oily appearance often described as 'scotch and soda' looking. Hessonite is found in Sri Lanka, Canada, Tanzania and Madagascar. Tsavolite is coloured by vanadium. Hardness is 7.25, SG is 3.57 to 3.65, RI is 1.739 to 1.744, and dispersion is 0.027. Dark green tsavolite may show broad vanadium bands at 430nm and 610nm. Inclusions may be tiny to large apatite crystals, fingerprint inclusions composed of tiny whitish dots, and groups of straight golden fibres. It is found in Tsavo National Park in Kenya.

Hydrogrossular garnet has the formula $H_2O \cdot Ca_3Al_2(SiO_4)_3$. It is semi-opaque polycrystalline massive green, white and pink sometimes known as 'Transvaal jade'. Pink coloured by manganese has RI 1.70 to 1.72 and SG 3.36 to 3.42; green coloured by chromium has RI 1.71 to 1.73 and SG 3.41 to 3.55. It shows strong orange/yellow under X-rays. Hardness is 7.25. Green is from Transvaal in South Africa, and Pakistan. White is found in Myanmar with jadeite. Inclusions are black pepperish. Spectrum shows a cut-off below 460nm.

Andradite is calcium iron silicate – $Ca_3Fe_2(SiO_4)_3$. The commercially significant transparent bright green to yellow-green variety coloured by chromium is known as *demantoid*. Hardness is 6.5; SG is 3.82 to 3.85, averaging 3.84; RI is 1.89; and the dispersion is 0.057 greater than that of diamond. Demantoid shows an absorption spectrum with a cut-off at 443nm, a doublet in the deep red at 701nm and weaker lines at 693nm, 640nm and 621nm. It exhibits moderate magnetism. Characteristic inclusion is byssolite fibre bundles like a horsetail; in combination with negative RI, this inclusion is diagnostic. It is from the Ural Mountains of Russia. It is the most pure garnet with less mixture than others. There is also a transparent yellow andradite and black *melanite* which is opaque and comes from Italy.

Tourmaline

Tourmaline is a complex boro-silicate of aluminium, with the general formula $XY_3Al_6(BO_3)_3(SiO_3)_6(OH,F)_4$, where X is usually an alkali metal – sodium, potassium or calcium – and Y is a metal – iron, lithium, magnesium, manganese or aluminium. It forms the most complex gemmological isomorphous series. Most gem varieties are from the series member known as *elbaite*, which is $Na(Li,Al)_3Al_6(BO_3)_3(SiO_3)_6(OH,F)_4$.

It crystallizes in the trigonal system with prismatic habit. Prisms are usually long, vertically striated, and basically three-sided, giving a rounded triangular outline in cross-section. Crystals are hemimorphic terminated with pyramid faces, which differ in angle from the terminations at the opposite end. Very short prisms have tabular habit. Euhedral crystals have no centre of symmetry and no plane of symmetry across the c axis.

Tourmaline is found in almost all colours. Varieties: red is known as *rubellite*; blue is known as *indicolite*; brown and yellow are *dravite*; reddish-violet is *siberite*; black opaque is *schorl*; colourless is *achroite*; and green is *verdelite*. Only rubellite and indicolite are commonly used names, the others are typically referred to as tourmaline with colour prefix. A teal blue from Paraìba, in Brazil, is known as *paraìba tourmaline*. A bright green stone owing its colour to chromium and/or vanadium is from Tanzania. It is red under the Chelsea filter and known as *chrome tourmaline*. Parti-coloured stones exhibiting two or more colours are also fairly common. Red centre with green rind is known as *watermelon tourmaline*. This colour combination may also be found as parti-coloured with red and green at either end and a colourless centre. Other parti-colour combinations include: deep blue or black with multiple layers of blue, green or pink. Chatoyant stones are also known.

Fracture is conchoidal to uneven, lustre is vitreous, and there is no cleavage. Constants are variable by colour, but in general most gem tourmaline is: SG 3.01 to 3.11 averaging 3.07, RI 1.622–1.640, with a birefringence of 0.018. Uniaxial negative. Dispersion is 0.017. Hardness averages 7 to 7.5. Diaphaneity varies from transparent through the full range to opaque. Its RI and high birefringence can identify most tourmaline if proper refractometer readings are taken.

Tourmaline is found in granites, granitic pegmatites, metamorphic gneisses and schists and in zones of contact metamorphism. It is typically associated with quartz, mica, corundum, topaz, spodumene and others, often as water-worn pebbles. Brazil is known for red, green, yellow, blue and parti-coloured. Russia is known for red, green and blue. Other sources are: Sri Lanka, Madagascar, California and Myanmar.

Tourmaline is piezoelectric and pyroelectric, meaning it becomes electrically charged when subjected to pressure in the direction of the vertical crystallographic axis and in response to heat in the range of 100°C. Iron-rich stones, like black tourmaline, do not exhibit these effects. Inclusions are usually thread-like cavities either as individuals or in mesh-like patterns. These are liquid-filled tubes or two-phase inclusions also containing a gas bubble. Also seen are flat films that appear as patches and crystalline inclusions of mica, quartz, tourmaline and zircon. Dichroism is strong in darker-coloured stones and usually shows as light and dark colours of the same hue. The colour in vibrations across the vertical axis is deeper. Tourmaline is typically cut as step, mixed and cabochon. More included material is often cut as beads. Heat treatment is usually undertaken to lighten the colour: dark Brazilian green and blue can be appealingly lightened. Irradiation can alter colour: colourless, very pale pink, green or blue can produce striking pink to red to purple stones. It can also turn some yellows to orange.

The absorption spectrum is weak but, if seen, green tourmaline may show: red absorption up to 640nm, yellow and green transmission, a narrow band centred at 498nm due to ferrous iron, and a weaker band at 468nm. Some green and blue show a band in the violet at 415nm. Red, pink and reddish-violet tourmalines show a broad absorption of the green centred on 525nm and there may be narrow lines at 537nm, 458nm and 450nm.

Tourmaline is generally inert to LWUV. Under SWUV, yellow

sometimes exhibits a very weak glow of indistinguishable colour; red and pink may glow blue to violet; and yellow, brown and green Tanzanian stones are sometimes a fairly strong yellow.

Peridot

This is silicate of magnesium and iron, with the chemical formula $(Mg,Fe)_2SiO_4$. It is a member of the olivine isomorphous series ranging from *forsterite* – magnesium silicate – to *fayalite* – iron silicate. Peridot is nearer to the magnesium end of the continuum, but contains 12–15% iron and this is the source of its green colour, making it an idiochromatic stone. Typically the colour is olive green to bright yellow-green. Stones containing chromium are bright green, as are stones with traces of nickel. With increasing amounts of iron colour becomes dark and is not commercial as a gemstone. Orange peridot has been reported. Chatoyancy and asterism are known but extremely rare.

It crystallizes in the orthorhombic system. Habit is prismatic and crystals are often flattened and vertically striated. Euhedral crystals are rare; it is most often found as water-worn pebbles. Hardness is 6.5, SG is 3.34, RI is 1.65–1.69 with a birefringence of 0.036. Biaxial positive. Lustre is vitreous to oily. There are two directions of cleavage, one of which may be rather distinct. The more prominent is parallel to the brachy axis and the less prominent one is parallel to the macro axis. There is no cleavage parallel to the basal pinacoid. Dispersion is 0.020. Trichroism is three shades of green: green, paler green to colourless, and yellowish-green. It is weak and not easy to see in pale stones.

It is found mainly in silica-poor igneous rock and in metamorphosed limestone. The Island of St John in the Red Sea has been known since ancient times as being a source of fine but small stones. Myanmar has large fine stones. Other localities are: Arizona, Hawaii, Brazil, China, South Africa and Norway.

The diagnostic feature is the *lily pad* inclusion, a disc-like formation consisting of residual liquid in a flat cavity left after a tiny chromite crystal crystallized out at the centre. Myanmar stones show rectangular biotite mica flakes. Hawaiian stones show oval- to pear-shaped drops of glass that can resemble bubbles. They are often seen doubled owing to the high birefringence of peridot and this separates them from the bubbles found in glass.

Peridot is typically cut as modified brilliants, step and mixed cuts, and fancy shapes. The ferrous iron content produces three bands in the blue: a broad band at 453nm, a narrow band at 473nm, and a band at 493nm. There may also be weaker lines at 529nm in the green, and 653nm in the orange. Peridot is inert to UV and X-rays. Peridot is rather easily identified by its RI and high birefringence.

Zircon

This is silicate of zirconium – $ZrSiO_4$ – in which some of the zirconium is replaced by iron and smaller amounts by thorium and uranium. It crystallizes in the tetragonal system. Its habit is a 4-sided prism of square cross-section terminated at each end by pyramids or by a combination of pyramids in differing inclination. Short crystals may seem pseudo-octahedral. Twinning is rare and is geniculate or knee-shaped.

Varieties: yellow, orange, red, brown and green. The popular colourless, blue and golden-yellow stones are all heat-treated brown stones. If the heating is carried out in an oxidizing atmosphere the colours produced are golden and colourless, and if treated in a reducing atmosphere blue is produced. Chatoyant stones are known.

Zircon is a metamict mineral in which the crystal lattice breaks down over time in response to the effects of radioactive elements. Zircon contains traces of radioactive uranium and thorium. Over time the decay of these elements bombards the crystal lattice with alpha particles that degenerate the lattice and leave the crystal amorphous, as amorphous silica and zirconia, or approaching amorphous. As crystal lattice breakdown is not rapid and it occurs on a continuum, zircons can be found in all stages of degeneration. In general, those that remain crystalline in which the process has not yet begun are termed *high type*, those that are fully amorphous are termed *low type*, and those in intermediate states of degeneration are termed *intermediate*. The constants for the low type are lower than those of the high type, with intermediates in between.

High-type zircon has hardness of 7 to 7.5, cleavage is negligible, and it is brittle. Fracture is conchoidal. Lustre may be sub-adamantine, but sometimes resinous. High-type RI is 1.925–1.984 with a birefringence of 0.059. Uniaxial positive. SG is 4.60 to 4.80. Dichroism is weak in all except blue stones: deep sky blue and colourless to yellowish-grey.

Dispersion is 0.038. Zircon exhibits a diagnostic strong, typically 10–14 band, spectrum of fine sharp lines due to traces of uranium. There is considerable variation in the number of bands seen in any one stone, but the strongest is at 653.5nm in the red and there may be up to 40 other lines. Colourless, blue and golden-yellow may only show the line at 653.5nm and it may be faint.

Low-type zircons have an SG of 3.90 to 4.20, hardness of 6.5, and RI may be as low as 1.78. Dispersion remains 0.038. There is no birefringence and no dichroism. Absorption spectrum shows a fuzzy line at 653.5nm. Colour is typically green; brown and orange are also known. Intermediate types have SG 4.20 to 4.60 and average RI 1.84–1.85 that ranges higher and lower with birefringence that also ranges higher and lower. Heat treatment to 1450°C will often return metamict zircons to the high type because the energy supplied as heat increases the vibrations and mobility of the atoms, causing the dissociated silica and zirconia to rearrange back into the crystalline state. This increases SG, RI and birefringence. Metamict zircons show characteristic inclusions: tension fissures, parallel stripes, healing fissures and ilmenite in fractures.

They are most often cut as modified brilliants, step and fancy cuts. Luminescence is variable under UV: some are inert and some fluoresce mustard yellow of variable intensity. Zircon should be straightforward to identify with its extremely high birefringence as doubling of the back facets should be easily seen. Zircon is a common accessory constituent in igneous rocks and the most important sources of zircon are Vietnam and Thailand. Gem-quality stones are known from Sri Lanka and Myanmar as well as Canada, Norway, France and Australia.

Chrysoberyl

This is double oxide of beryllium and aluminium, with the chemical formula $BeAl_2O_4$. It crystallizes in the orthorhombic system. Habit is prismatic crystals flattened parallel to one pair of faces. Twins are cyclic, known as trillings, which gives a pseudo-hexagonal appearance to the crystal. Much is found as water-worn pebbles.

There are several varieties. Transparent stones are pale yellow to golden-yellow to greenish-yellow, yellowish-green to green, and brown. Alexandrite is the transparent colour change variety. Chatoyant stones, known as *cymophane*, are translucent to opaque, and yellow to greenish-yellow to brownish-yellow. Alexandrite is occasionally chatoyant. Star chrysoberyl is rare.

Cat's eye stones are due to tiny needle inclusions of exsolved rutile running parallel to the c axis. Usage of the term 'cat's eye' with no qualifier is restricted to chrysoberyl. Other stones such as quartz cat's eye must be prefixed with the mineral species. Translucent stones of honey-yellow colour with a straight sharp eye are most prized. Typically the eye is more sharply defined than it is in other minerals.

There is small-scale replacement of aluminium with chromium oxide that gives alexandrite its colour. Alexandrite is green in daylight or fluorescent light and red in incandescent light. The colour is evenly balanced, meaning it transmits evenly between the red and the green, thus the nature of the light determines its colour. Alexandrite is one of the most expensive stones. Colombian emerald green to fine ruby red is the ideal colour change. Pleochroism is weak in yellow stones. In brown stones it is three shades of brown. In alexandrite in fluorescent or daylight it is pink, yellowish and bluish-green; and in incandescent light it is red, orange and green. Spectrum varies with vibration direction. In the vibration direction of the green ray: narrow line at 680.5nm and a weaker line at 678.5nm, possibly weak lines at 665nm and 649nm, a broad absorption from 555nm to 640nm, and absorption of the violet below 470nm. In the vibration direction of the red ray: the 678.5nm line is stronger than the 680.5nm line, lines at 655nm and 645nm, broad absorption between 540nm and 605nm, a line at 472nm and absorption of the violet from 460nm.

Inclusions of transparent chrysoberyl are: two-phase inclusions – cavities filled with a liquid and a gas bubble, flat liquid cavities, long tubes, and solid crystal inclusions of mica, actinolite and goethite needles, quartz and apatite. An unusual feather is a stepped twin plane. Inclusions are typically feathers in alexandrite, but any of the other inclusions are possible. Transparent yellow and brown due to ferric iron show a strong band at 444nm.

There are three directions of cleavage, only one of which is distinct. Fracture is conchoidal but sometimes uneven. Hardness is 8.5. Lustre is vitreous. RI is 1.74–1.75, with a birefringence of 0.009. Biaxial positive. SG is 3.71 to 3.72. Dispersion is 0.015. Iron-rich yellow, brown and green transparent chrysoberyl is inert to UV. Some pale greenish-yellow may be faint green under SW. Alexandrite shows a weak red glow to UV and a dim red to X-rays. The usual cut is mixed cut, but more rarely it is step cut. Chatoyant stones are cabochons.

Chrysoberyl of all types is known from Sri Lanka and Brazil. Sri Lanka is the source of the finest cat's eyes. Transparent stones are found in pegmatites in Myanmar and Mozambique. Historically, alexandrite was from the Ural Mountains of Russia, but today the most important deposit is in Minas Gerais, Brazil. Alexandrite from Myanmar and Tanzania is associated with emerald and phenakite in mica schist. India is a source of chatoyant stones.

True synthetic alexandrite is produced. Typically the colour change is more pronounced than it is with natural stones. Chromium lines are seen in the spectrum and there is a strong red fluorescence under UV. Czochralski pulled stones show gas bubbles and fine curved growth striae. Flux stones show feathers of flux and veil-like inclusions, chains of solidified flux droplets, metallic platelets, and straight growth lines. Gas bubbles and colour swirls are found in other stones. Alexandrite is also imitated by synthetic colour-change corundum, but this has a different RI, SG and spectra. Also, the colours are different: greyish-blue in daylight and strong purple in incandescent light; it never has the intense green of natural alexandrite. Colour-change glass, usually of pale hues, shows colours closer to the synthetic colour-change corundum than to alexandrite. Some colour-change garnets can look very similar to alexandrite, but are isotropic.

Quartz

Silicon dioxide, SiO_2, or silica, is found abundantly all over the world. Oxygen constitutes 46.6% of the Earth's crust and silicon 27.7%, therefore it is a common mineral but fine gem material is a very small part. Quartz gems are divided into two general groups: crystalline and multi-crystalline.

Crystalline quartz

Quartz crystallizes in the trigonal system with prismatic habit as a hexagonal prism terminated at one end, rarely at both ends, by positive and negative rhombohedra, sometimes with other modifying forms.

Positive rhombohedra usually predominate. Prism faces are striated horizontally and twinning is almost universal.

Hardness is 7. SG is 2.65, RI is a very constant 1.544–1.553, with a birefringence of 0.009. Uniaxial positive. Dispersion is 0.013. Fracture is conchoidal, lustre is vitreous, and dichroism is weak. Quartz is inert to UV.

Clear colourless quartz, known also as *rock crystal*, may contain two-phase inclusions, often gas or liquid CO_2, with various minerals. Fine crystal group specimens are known from the Swiss and French Alps. The most important source is Brazil and also Madagascar, Japan and Myanmar. There are various varieties of included rock crystal: *sagenitic quartz* – included needles resembling a fish net; *rutilated quartz* – red or golden rutile needle inclusions, alternatively known as Venus hair stone; *Thetis hair stone* – long slender inclusions of dark tourmaline or actinolite; and *dendritic quartz* – dendrites of green chlorite, goethite or red to orange hematite in colourless quartz. *Rainbow quartz* exhibits cracks showing interference colours; sometimes these cracks are filled with dye.

Brown quartz and *smoky quartz*, a grey-brown colour with smoky tinge, are coloured by a colour centre involving an aluminium impurity. Natural radiation ejects and traps one of a pair of electrons, leaving a hole colour centre that absorbs light and yields the brown colour. Typical inclusions are two-phase negative crystal cavities with CO_2 as the liquid. The most important source is the Swiss Alps. No distinct dichroism or spectrum.

Citrine, yellow quartz, owes its colour to a trace of ferric iron. It varies in colour from light golden-yellow to reddish-yellow. The best is from Brazil and there is perceptible dichroism.

Amethyst, quartz ranging in colour from faint mauve to violet-purple, owes its colour to radiation-induced colour centres associated with an iron impurity. Due to twinning, much amethyst is parti-coloured and shows angular zones of colour. Inclusions are: feathers of negative cavities and marks like a fingerprint, known as tiger stripes, ascribed to rhombohedral twinning. It shows slight to distinct dichroism: bluish-purple and reddish-purple. It is reddish under the Chelsea filter. Spectrum shows a wide absorption of varying intensity in the yellow-green at 520–550nm, but it is not distinctive. Some exhibits a blue glow to SWUV. Historically, the source is Russia, associated with beryl and topaz. Sources include: Brazil, United States, Canada, Myanmar and Madagascar.

Rose quartz is pink and nearly always cloudy with fissures. Much of it contains microscopic needles of rutile that produce a star effect when oriented correctly. The star is seen in transmitted light, an effect known as *diasterism*. There is fairly strong dichroism in the deeper-coloured material and no distinct absorption spectrum. Rose quartz has crystal properties but massive habit.

Cat's eye quartz owes its chatoyancy to a multitude of slender crystals or channels parallel to the principal axis. It is honey-yellow to brownish to grey-green. Sri Lanka is the source of the finest stones.

Other varieties include: white – *milky quartz*; green transparent – *prasiolite*; and transparent with the colours of amethyst and citrine side by side – *ametrine*. When amethyst is heated it usually changes to yellow citrine. This can be distinguished from natural-coloured yellow material by its lack of dichroism. Some amethyst turns greyish-green (prasiolite). Smoky quartz when heated turns through various hues to colourless, while brown quartz may become reddish.

Crystalline quartz is cut in modified brilliant, step and mixed cut styles. Included stones, cat's eyes and star stones are cut as cabochons. Quartz is piezoelectric and pyroelectric, meaning it becomes electrically charged when subjected to mechanical stress such as friction, or heat. Quartz is found in open cavities in all classes of rocks. Amethyst occurs in geodes in volcanic rocks. The most important locations are: Brazil, Sri Lanka, Russia, Madagascar, Japan and Scotland for colourless, citrine and amethyst. Star quartz is found in Brazil, Madagascar and India. Synthetic quartz is produced primarily for technical uses and is typically untwined, and can usually be distinguished from natural quartz by its interference figure.

Multi-crystalline quartz

There are four general varieties of multi-crystalline quartz: (1) quartzite; (2) cryptocrystalline – chalcedony; (3) fibrous cryptocrystalline pseudomorphs; and (4) granular microcrystalline – jasper.

Quartzite is a rock consisting of a granular interlocking mass of quartz crystals with irregular boundaries. It is formed metamorphically from pre-existing rocks and also formed by silica sedimentation of sand or sandstone. The important variety is *aventurine*, containing small crystals of mica or an iron mineral causing sheen. It may be creamy white to reddish brown, but the most prized has inclusions of chrome green fuchsite mica. It is used for ornamental purposes, but it may be cut into beads. It has the RI of crystalline quartz but shows only a vague shadow edge at 1.55 owing to the mass of disoriented crystals. The SG is 2.64 to 2.69 due to inhomogeneity. Green aventurine shows red through the Chelsea filter and a chromium spectrum, but the lines in the red are vague and the absorption of the yellow-green is not pronounced. It is greyish-green under UV and the other colours are inert.

Chalcedony is a general name applied to any fine-grained quartz with cryptocrystalline microstructure. Hardness averages 6.5 and SG ranges from 2.57 to 2.64 due to porosity and water, and impurities. RI is between 1.530–1.539, but most commonly it is 1.534–1.538, with a birefringence of 0.004. This is 'form birefringence', which is shown by materials of one RI embedded in another material of a different RI. Fracture is splintery and lustre is waxy. UV response is usually bluish-white but material containing uranium, such as that from Wyoming, is bright yellowish-green. Chalcedony is more responsive to SW; many are inert to LW. Chalcedony is somewhat porous and therefore can be stained in various colours. Chalcedony is very widespread and the main locations are Brazil, Madagascar and Uruguay.

Chrysoprase is a prized green chalcedony. The colour is due to the presence of disseminated particles of a hydrated nickel silicate. It varies in colour from apple green to dingy greenish-yellow. It is simulated by green agate stained by chromium salts, which shows red under the Chelsea filter, whereas chrysoprase shows green. The rare chromium chalcedony found in Zimbabwe, known as *mtorolite*, shows red under the Chelsea filter and a sharp one-band spectrum in the red.

Other varieties of chalcedony include: *prase* – a more translucent green variety coloured as a result of fibrous hornblende or disseminated chlorite; *sard* – uniform brown to dark brown; *carnelian*, or cornelian, is red to reddish-brown or flesh-coloured. It borders sard in colour at the reddish-brown end and distinction is arbitrary in these cases; and *myrickite* – translucent white with streaks and clouds of bright red or pink cinnabar.

Agate is a variety of chalcedony that displays distinct banding with layers in differing colour and degrees of translucency. Natural colours are green, yellow, red, reddish-brown, white, bluish-white and others. It often shows fluorescence in bands or patches. Heating alters brown to red. Oiling improves translucency. Many types have individual names:

onyx has white alternating with black layers; *sardonyx* has layered reddish or brownish alternating with white; *moss agate*, also known as mocha stone, is pale bluish, grey or yellow with dendritic inclusions of black, green, yellow-brown or red minerals; and *scenic agate* has inclusions giving the appearance of a landscape. Agate was historically found in Egypt. Current important sources are Brazil and Uruguay.

A pseudomorph exists when a mineral occurs in a habit not typical of its species by replacing or changing from another mineral or material. The important fibrous cryptocrystalline pseudomorph of quartz is *Tiger's eye*, often termed *crocidolite*, a translucent to opaque yellow to golden-brown stone that shows a series of lustrous yellow bands alternating with brown bands. It is usually cut in flat plates or as beads and cabochons. The original material of the veins was a blue asbestiform variety of riebeckite known as crocidolite that altered by decomposition into silica. The fibrous structure is usually stained by hydrous iron oxides, which give the colour. There is also agatized wood, agatized bone, and agatized coral. Also, seaweed and small mollusc shells may be fossilized by silica.

Jasper consists of a heterogeneous mass of microcrystalline quartz containing significant amounts of other materials. It is usually dark brownish-red, but it can be yellow or black. Finely divided hematite gives the red colour and goethite gives the brown and yellow colours. When banding is present it is more often planar rather than concentric like agate. Jasper occurs as extensive beds of sedimentary or metamorphic origin. SG varies between 2.58 and 2.91, but most are below 2.65. Hardness is slightly below 7. The RI is 1.54 and there is no luminescence or absorption spectrum. It is sometimes named by patterns: ribbon jasper, orbicular jasper, and brecciated jasper. *Plasma* is microgranular or microfibrous quartz in various shades of green due to included chlorite. Plasma often contains spots of iron oxide or red jasper, and this is called *bloodstone*.

Opal

Opal is amorphous solid silica gel with varying amounts of water, and the chemical formula $SiO_2 \cdot nH_2O$. Gem-quality material is usually 6–10% water. It is found filling nodules as veins or seams in rock, or as botryoidal masses in cavities and as stalagmitic deposits. Precious opal must display

play of colour. Its structure is a stacking of tiny spheres of silica with regular voids in between that produce a rainbow-coloured effect where the colours seen at each location progress through a partial series of hues when the stone, light source or the observer moves. Opal is structured in patches containing uniformly sized spheres, with the size varying from patch to patch. The mechanism by which they produce colours is discussed in Lesson 6.

The four varieties of gem opal include: (1) *white opal* – this is the most common precious opal with play of colour in a light or milky white body colour; (2) *black opal* – the rarest and most valuable opal with play of colour on a dark grey, blue or black body colour; (3) *fire opal* – translucent reddish-orange to red variety which may or may not display play of colour; and (4) *water opal* – play of colour in a clear transparent stone, also known as jelly opal. Common opal is opal with no play of colour. It can be translucent to opaque and green, cherry-red, rose pink and yellow. Common opal in association with precious opal is known as *potch*. Boulder opal is a natural, not assembled, stone with a layer of precious opal on top of matrix.

The SG of opal is 1.98 to 2.20 and averages 2.10. Fracture is typically conchoidal to irregular, but may also be smooth. Lustre is vitreous. There is no cleavage. The hardness is 5.5 to 6.5 and RI averages 1.45. It is cabochon cut and often made into composite stones such as doublets and triplets. Fire opal has an SG of 2.0 and an RI of 1.40. It is modified brilliant or mixed cut. Precious opal of light body colour shows variable fluorescence to UV ranging from white, bluish, brownish or greenish with persistent phosphorescence. Black opals are usually inert. Fire opals show greenish-brown.

The historically important source is Czerwenitza in Czechoslovakia. Black opal is best known from Lightning Ridge, Australia. Also important are White Cliffs and Coober Pedy in Australia. Fire opal is known in Mexico. Opal is also found in Brazil, the United States, Zimbabwe and South Africa.

Impregnation with oil or plastic improves play of colour and prevents or disguises cracking, known as crazing, caused by the water drying out. Impregnation with black plastic imitates black opal. Impregnated opals typically have lower SG and RI. The appearance of black opal can also be achieved by smoke impregnation or dyeing with silver nitrate or sugar carbonized by acid. In these cases the black colour is in grains, which is different from the structure of natural black opal.

Gilson-created opal has a slight compositional difference as very small amounts of ZrO_2 have been detected; therefore, although it is often termed a synthetic, it is actually an imitation. Gilson opals are produced to imitate both white and black precious opal. The process is secret, but the opals are thought to be composed of a silicon ester solution. These silica particles, which are similar to the cristobalite spheres that compose natural opal, are allowed to settle into closely packed arrays through a process of sedimentation. After settling, any excess water is removed. This produces highly porous and brittle materials that are impregnated and stabilized with opaline silica. Difficult to distinguish with the naked eye, but magnification reveals that the structure of the iridescent patches is markedly different. Something that looks like crumpled tinsel paper or triangular patches of iridescent colour are common observations. Natural opal exhibits completely flat planes with a striated satiny finish. Also, the Gilson colour patches are columnar when viewed from the side, and from the underside of a cabochon these colour patches will often appear to be all the same colour. Gilson black opal may have long ribbon-like streamers of colour that are unlike natural structure. The colour segments of both black and white Gilson opals have been described as having a 'lizard skin' appearance. Gilson opals show only weak fluorescence.

A plastic-simulated opal is sold as polished cabochons that show play of colour with pinfire and flash effects in various colours. The simulant is marketed under the trade name of Opalite. Technically, it is an assembled stone as a layer of different plastic overlies the base. The outer layer may be an acrylic resin coated over the polystyrene imitation opal to protect it, and when it is immersed the clear plastic shell can be seen. Hardness is only 2.5. The RI is 1.50 to 1.51 and SG is 1.20. Under incandescent light, Opalite is translucent milky bluish-white and appears pinkish-orange in transmitted light. The colour patches also show the 'lizard skin' appearance. Under magnification it is similar to natural opal with spheres of polymer instead of spheres of silica. The play of colour is good so it may be convincing. It fluoresces strong blue-white to LW with no phosphorescence and is hydrophobic, meaning not wetted by water.

Turquoise

Turquoise is hydrous copper aluminium phosphate with some iron, and has the chemical formula $CuAl_6(PO_4)_4(OH)_8 5H_2O$. Some alumina is replaced by ferric oxide. It is a cryptocrystalline aggregate with the crystallites so small it may be considered amorphous. Only at Lynch, in Virginia, was it found in distinct triclinic crystals. The most desired colour is robin's egg blue. The finest and most valuable is Iranian, referred to commonly as Persian turquoise. Egyptian has a tendency to the greenish hues. American material is paler in colour, more porous and chalky. Blue colours are due to copper and greenish colours due to iron. Turquoise is also found in Chile, Russia, Australia and China.

Hardness is 5.5 to 6. Lustre is waxy. It is almost opaque. The mean RI is 1.62. SG is 2.6 to 2.9, with American material at the lower end and Iranian material at the higher end. It is almost always cut as cabochons or used for inlay work. The spectrum, seen in reflected light, shows a vague band at 460nm and a line at 432nm forming a distinctive pattern. The line at 420nm is too far in the violet to be seen. Under LWUV it varies from greenish-yellow to bright blue and is inert under SW and X-rays.

Turquoise is mostly found in arid regions and occurs as encrustations, nodules, botryoidal groups and veins. Blue turquoise may turn greenish if exposed to perspiration or grease. American turquoise is softer and much of it is bonded with plastic. Also, pressing together a precipitate of aluminium phosphate coloured blue by copper oleate makes a pressed simulant. The SG of bonded material is 2.4, but the reading must be taken quickly as soaking will increase SG to 2.60. RI is about 1.56. The characteristic appearance under the microscope is of white specks against a uniform blue background. A simple test for pressed and bonded turquoises is to place a drop of hydrochloric acid on the back of the cabochon. The acid turns yellowish-green and will stain a Kleenex when wiped. Some imitations exhibit a strong blue under SWUV.

Gilson produces a turquoise simulant. Its composition has never been publicized, but it is probably a ceramic made by precipitation, grinding and pressing. Natural turquoise contains iron, and since there is no trace of iron in the simulant it is known that the starting materials are not ground natural pieces. It is available in medium to darker blues, with or without matrix, and is colour stable. The SG averages 2.74 and the RI averages 1.60. Microscopic identification is required: it shows a mass of angular dark blue particles against a whitish groundmass at 30–40X.

Turquoise is frequently oiled or waxed to improve colour. This is commercially acceptable. Also acceptable is plastic impregnation. Turquoise is often stained blue and a drop of ammonia on the back of the stone can detect this, as it will appear green. Imitations of turquoise are made in glass, enamel, stained chalcedony and, more rarely, porcelain. Stained chalcedony is more translucent and has an RI of 1.53 and SG of 2.63. Glass has an SG near 3.3 and usually shows small bubbles below the surface or as pit marks on the surface. Porcelain imitations have a china lustre and SG 2.3 to 2.4.

Spodumene

Spodumene is lithium aluminium silicate, $LiAl(SiO_3)_2$, a member of the pyroxene group. It is monoclinic and crystals are usually flattened prisms heavily vertically striated with irregular capping. Varieties are transparent: bluish-pink or lilac-pink – *kunzite*; emerald green coloured by chromium – *hiddenite*; and yellow, yellowish-green and pale green coloured by iron. Blue has also been reported, as has chatoyant kunzite. Irradiated green colours rapidly fade.

Cleavage is prismatic in two directions at almost 90° and fracture is uneven. Hardness is 7 and lustre is vitreous. RI is 1.660–1.675 with birefringence of 0.015. Biaxial positive. SG is 3.17 to 3.19. Dispersion is 0.017. Trichroism is strong in kunzite: violet, deep violet and colourless; and hiddenite: bluish-green, emerald green and yellowish-green. Kunzite is noted for golden-pink or orange fluorescence under LW with a weaker response under SW. Yellow and green shows weak orange under LWUV and very weak orange under SWUV. Both exhibit strong and persistent phosphorescence after X-ray stimulation. Hiddenite is inert to UV but glows slight orange with some phosphorescence to X-rays.

Spectrum is not diagnostic for kunzite. Hiddenite shows a chromium spectrum with a doublet at 690.5nm and 686.0nm, weaker lines in the red and orange at 669nm and 646nm, broad absorption centred near 620nm, and general absorption of the violet. Yellow and other green stones show a well-defined band in the blue at 437.5nm and a narrow band at 433nm, both due to iron. Liquid inclusions and growth tubes are characteristic inclusions. RI readings separate kunzite from other genuine pink stones as well as synthetic pink spinel and pink glass.

Primary localities for spodumene are North and South America, Madagascar and Myanmar. Hiddenite was discovered in North Carolina in 1829. Kunzite is found in Afghanistan and California. Yellow spodumene is found in Brazil.

Jades

The word 'jade' refers to two mineral species: jadeite and nephrite. They are similar in appearance but are otherwise unrelated. Both are cut as cabochons, beads, bracelets or ornamental carvings. Nephrite is much more common and Canadian nephrite is sold worldwide. SG is the best test for a loose stone.

Nephrite is a silicate of magnesium and calcium with some ferrous iron, $Ca_2(Mg,Fe)_5(OH)_2(SiO_4O_{11})_2$. It is monoclinic and almost cryptocrystalline in character. Its habit is compact, tough, fibrous masses. It is translucent to opaque and usually green, which deepens in colour with an increase of iron. With more magnesium its colour is creamy, known as mutton fat jade. Brownish is caused by oxidation of iron content. Yellow or greyish-brown is known as buried jade, as its colour and perhaps composition have altered due to a long burial. Cleavage is lengthwise and not easily developed, but not ever seen in the multi-crystalline mass. Fracture is splintery. Hardness is 6.5, mean RI is 1.62, and the RI of individual fibres varies from 1.600–1.627 to 1.614–1.641, with a birefringence of 0.027. SG is 2.90 to 3.02, averaging 3.00. Nephrite shows a very indistinct spectrum of a doublet in the red about 689nm, vague bands at 498nm and 460nm, and a sharp line at 509nm. It is inert under UV and green under the Chelsea filter.

Nephrite is most known from Canada, but also the United States, New Zealand and Russia. It is found *in situ* or as alluvial boulders and pebbles. It is usually a product of metamorphic action and is found typically with gneiss, schist, serpentine or metamorphosed limestone.

Jadeite is sodium aluminium silicate, $NaAl(SiO_3)_2$. It is one of the pyroxene group of minerals. It is an aggregate of interlocking crystals of the monoclinic system with a habit more granular than the fibrous nephrite. It is translucent to opaque and the colour range is much wider than for nephrite. Jadeite is found in an extensive range of green as well as white through pink, brown, red, orange, yellow, mauve, blue, violet and black. Also, it is found in mottled green and white. Mauve colour has been attributed to manganese, fine green colour to chromium, and paler green to iron.

Cleavage is imperfect and fracture is splintery. Hardness is 7, RI for an individual crystal is 1.654–1.667 with a BI of 0.013, but only a vague edge is seen on the refractometer at 1.66. SG is 3.30 to 3.36, averaging 3.34. *Chloromelanite* is a jadeite containing a considerable amount of iron oxides; it is dark green to black in colour with a higher SG of 3.5. Spectrum shows a strong line in the violet at 437nm, weaker bands at 450nm and 433nm, and maybe a band at 495nm. The 437nm is diagnostic but can be difficult to see in rich green material coloured by chromium, as absorption of the violet tends to mask it. Chromium-rich green stones show a strong band at 691.5nm, which is an unresolved doublet of 694nm and 689nm, and weaker lines at 655nm and 630nm. Under LW, the pale colours glow whiteish and the darker colours are inert. Most are inert to SW.

Pale jadeite can be stained a fine green Imperial jade colour; this colour can be seen concentrated along cracks and the spectrum shows two broad bands in the red which are diagnostic for stained jadeite. It is prone to fading. Dyed mauve is also produced, but the colour is pronounced and appears unnatural and shows slight orange fluorescence to LWUV. It is less prone to fading. It is also assembled as a triplet with a white cabochon core cemented with green cement to a hollow top and the base closed by a third piece of jadeite. The spectrum is similar to the dyed material with one or two broad bands in the red. Plastic jadeite imitations have much lower SG.

Myanmar is the primary location for jadeite, where it is found as alluvial boulders and *in situ* in dykes of metamorphosed rock. Other locations include Russia, the United States and Japan.

Feldspars

Feldspars are aluminous silicates of potassium, sodium, calcium or barium, practically without magnesium or iron. It is an isomorphous group of minerals closely related chemically, but with considerable isomorphous replacement and with intermixture of monoclinic and triclinic crystal systems.

There are two subgroups: (1) potassium aluminium silicates, $K(AlSi_3O_8)$, which includes *orthoclase* feldspar which crystallizes in monoclinic system, and *microcline* feldspar which crystallizes in the triclinic system; and (2) sodium and calcium aluminium silicates, $(Na,Ca)(AlSi_3O_8)$, which are the *plagioclase* feldspars that crystallize in the triclinic system. In general, feldspar has two directions of easy cleavage nearly at right angles to one another. Hardness is 6, RI is 1.52–1.58, and SG is 2.5 to 2.7.

Orthoclase feldspar is transparent and colourless in its purest form and is not gemmologically significant. The most commercially important feldspar is *moonstone*, a mixture of orthoclase and plagioclase. It is near-colourless orthoclase that contains innumerable tiny inclusions of albite uniformly scattered throughout it. The nature of the albite inclusions and how they produce the fine blue sheen of moonstone is discussed in Lesson 6. Stones may show chatoyancy or asterism. Moonstone SG varies from 2.56 to 2.59, hardness is 6, and RI is 1.520–1.525, with BI of 0.005. Biaxial negative. Dispersion is 0.012. Lustre is vitreous and there is no characteristic spectrum. It is found in gem gravels in Sri Lanka and in India, Myanmar, Madagascar and the United States. It usually exhibits a bluish glow to LWUV and a weak orange glow to SWUV. Typical inclusions are stress cracks consisting of straight cracks parallel to the vertical axis, usually in pairs or multiples, and numerous branching cracks extending for a short distance in the direction of the b axis. These characteristically appear as what is known as the moonstone centipede inclusion and it is diagnostic. Yellow orthoclase is coloured by iron and is a collector stone. It may also be orange or green. Found in Madagascar.

Amazonite is the most commercial of the microcline feldspars. It is green to bluish-green and used for beads and cabochons. Microcline occurs in translucent to opaque well-formed crystals or as massive greyish-white, flesh-coloured, brown-red or green. Hardness is 6.5, SG is 2.56 to 2.58, and RI is 1.522–1.530, with BI of 0.008. Biaxial negative. There is no distinct spectrum and it is normally yellowish green under LW and inert under SW. It has two directions of good cleavage and is not suitable for carvings. The most important source is India, but it is also found in Brazil, Canada and Madagascar.

The plagioclase feldspars, also known as the soda-lime feldspars, form an isomorphous series from sodium-rich albite, $NaAlSi_3O_8$, to calcium-rich anorthite, $CaAl_2Si_2O_8$. The series includes in order of increasing calcium content: albite, oligoclase, andesine, labradorite, bytownite and anorthite. In general, plagioclase crystallizes as masses and not good crystals. The most important variety is *labradorite*. It is translucent greyish with iridescence known as labradorescence. This colour effect is discussed in Lesson 6. Labradorite has an SG of 2.69, and an RI of 1.560–1.568, with a BI of 0.008. There is no observed spectrum and it is inert to UV. Its notable location is Labrador, in Canada. It is also found in other Canadian locations as well as Ukraine, Russia and the

United States. Labradorite without the grey background colour, from Madagascar and elsewhere, is termed spectrolite and is sometimes marketed by the misnomer 'rainbow moonstone' – a term that should be avoided.

Other gemmologically significant feldspar varieties include: *sunstone* – an oligoclase sometimes called aventurine feldspar. It is rich golden or reddish-brown with specular reflections. SG is 2.62 to 2.65 and RI is near 1.54–1.55. It is inert to UV; *peristerite* – an oligoclase with blue sheen on a white, cream or brownish-pink body colour. SG is 2.617 and RI is near 1.54; and *bytownite* – a reddish or sometimes pale yellow faceted stone. SG is 2.739, RI is 1.56–1.57, and there is a band in the spectrum at 573nm.

LESSON 12

Synthetic Gemstones

Synthetic by definition is applied only to inorganic materials, which have essentially the same chemical composition, crystal structure and physical and optical properties as their natural counterparts, although small variations in composition as a result of impurities are accepted. Note by this definition that a synthetic stone must have a natural counterpart, even if this counterpart is not desirable as a gem. Organic gemstones, such as ivory, pearl and coral, are *simulated*, not synthesized, as they are the products of biological growth and the appropriate term for such material produced by human initiation, or under human control, would be *cultured*.

Synthesizing gemstones was not always pursued. Natural gemstones have a long history of desirability, but through most of those centuries gemstones were thought to have magical powers and thus attempts at synthesis would have been against this belief. It was only in the eighteenth and nineteenth centuries that scientists came to a better understanding of what minerals were and how they formed naturally in the earth. It was then that experimentation began. Today the synthetic gem industry produces billions of carats each year.

There have always been three major challenges. First, natural gemstones grow on a geological time scale over a period of years to millennia to aeons, but to be commercially viable synthetics must be induced to grow in a matter of hours, days or, at most, months. Second, gemstones grow under extremes of temperature and pressure, some of which is well above the limits of what can be created in a laboratory. And third, they must be produced at a fraction of the cost of the natural version and in crystals large enough for cutting in order to have value in the marketplace. Many gemstones are made solely for industrial or technical applications, such as synthetic rubies for lasers, synthetic diamonds for thermal management applications, and synthetic quartz for watches. The most valuable gemstones – diamond, ruby, emerald, sapphire – are also produced for the jewellery industry.

Different gemstones have different production techniques. In general there are two categorical types: *melts* and *solutions*. Melt techniques start with a mixture of chemical elements or compounds that will provide the composition of the desired gemstone. The mixture is melted without the aid of a flux or solution and then cooled under controlled conditions in order to yield large single crystals when solidified. This is a similar process to producing ice cubes in the freezer. Examples of melt techniques are: Verneuil flame fusion, Czochralski pulling technique, and skull melt. Overall it is a simple technique that yields rapid growth, but it cannot be used on certain gems, such as quartz, where the SiO_2 becomes so viscous that it forms a glass, and emerald, where the material melts incongruently, meaning it decomposes below the melting point. Solution techniques

Table 12. 1 Synthetic gemstone synthesis techniques

Melt techniques
Verneuil flame fusion
Czochralski pulling
Skull melt
Solution techniques
Flux
Flux transport
Flux reaction
Flux melt
Hydrothermal
Hydrothermal reaction
Hydrothermal transport
High-pressure solution
Others
High-pressure sublimation

yield growth by an increase in saturation, a reaction, or by transport. Examples are the various flux techniques involving solution growth in molten solvents – diamond is synthesized by this technique under extreme high pressure and with a molten metal as the flux; and hydrothermal techniques in which the growth is in water at high temperature and pressure. Other techniques include high-pressure sublimation, which produces synthetic moissanite. Each will be described and presented along with a diagram. Notes on the identification of synthetic stones are included.

Verneuil flame fusion – synthetic corundum

The first step towards synthesis was Joseph Gay-Lussac's 1817 report that aluminium oxide could be obtained by heating ammonium alum. Then, in 1837, American mineral engineer Marc A. A. Gaudin reported production of tiny hexagonal platelets of corundum using a downward-pointing flame. These could be modified into ruby if chromium salt was added; profuse bubbles characterized them. In 1877, French chemist Edmond Frémy and his assistant, C. Feil, published an important paper on ruby growth, and he became the first to produce small but clear red ruby crystals. In 1885, a number of rubies appeared on the jewellery market that were accepted as natural stones. Soon this was questioned, and it later transpired that these stones, known as *Geneva rubies*, had supposedly been made by the direct fusion of small fragments of real ruby. The stones had the physical and optical characteristics of natural ruby, but contained many gas bubbles, often irregularly arranged, giving the stone a cloudy effect. Many contained flaws as a result of the rapid cooling and some showed characteristic whirl striae like that seen in badly annealed glass. Due to their method of production, said to be a form of reconstruction, these stones were known as *reconstructed* rubies; however, research by Kurt Nassau has revealed these stones to be of notably low iron content and that a highly purified aluminium oxide was used in their production. This proves the starting material to be other than natural rubies and such stones are no longer referred to as reconstructed rubies.

After some years of research, the French chemist Auguste Verneuil, who served for sixteen years as Frémy's laboratory assistant after the death of Feil, discovered a method whereby true synthetic corundums of all colours and of large size could be made by a *flame fusion* process. He constructed a special form of inverted oxy-hydrogen blowpipe, referred to as his *chalumeau,* and in 1904 published a short account of his experiments. The difficulty for early synthesis experimentalists was that they lacked the ability to achieve the high temperature, more than 2050°C, needed to melt the alumina. Verneuil was able to overcome this with the oxy-hydrogen flame. Verneuil's is the best-known method for the production of synthetic corundum, as well as synthetic spinel, strontium titanate and synthetic rutile. With secondary annealing, star stones and imitation moonstone can be produced. Over 1000 million carats of synthetic corundum are produced annually by this method, and a great deal of this is for technical uses.

First and foremost is the preparation of a suitable feed powder as impurities have detrimental effects on crystal growth, such as undesirable colour or cracking. To avoid this, Verneuil started with ammonium alum, $(NH_4)Al(SO_4)_2 \cdot 12H_2O$, which he dissolved and distilled in water. The solution was filtered to remove all insoluble matter. Cooling caused crystals of alum to reform from the solution while soluble impurities remained in the solution. The procedure was repeated four or five times to achieve the needed purity. For ruby production, he repeated the procedure with chromium alum: $(NH_4)Cr(SO_4)_2 \cdot 12H_2O$. A mixture of the two alums was fired at a temperature of 1000–1200°C to decompose it into the required aluminium and chromium oxides: Al_2O_3 and Cr_2O_3. The mixed oxide was ground to a fine powder used for production.

The Verneuil apparatus is a vertical blowpipe consisting of two iron tubes, one of which is widened into a large chamber at the top. The lower and thinner end of this tube passes down the centre of the other tube into which it is tightly screwed, forming a gas-tight joint. In the upper chamber is a feed hopper with a fine mesh bottom. The finely powdered mixture of aluminium and chromium oxides is placed in the hopper. This is mechanically vibrated by the strike of a hammer giving a series of regular taps, 20 to 80 per minute, on an anvil head to release the powder in discrete amounts. Oxygen enters the blowpipe through a pipe at the upper end of the chamber and the oxygen current carries the released powder with it down the inner tube of the furnace. An outer annular tube carries hydrogen under pressure. The ends of the two tubes form a nozzle at which point the gases are ignited, producing a downward-pointing flame with an intense heat. The powder melts as it passes through the flame and falls on a ceramic pedestal called a *candle*. The

SYNTHETIC GEMSTONES

candle is on an iron rod connected to a screw adjustment so it may be raised and lowered. The fused alumina solidifies on the end of the candle, forming a small cone consisting of a number of crystals. The tip is kept in the hottest part of the flame by adjustment of the height of the candle and remains molten. By control of the rate of powder dropped and pressure of the gases, one of the crystals is induced to grow.

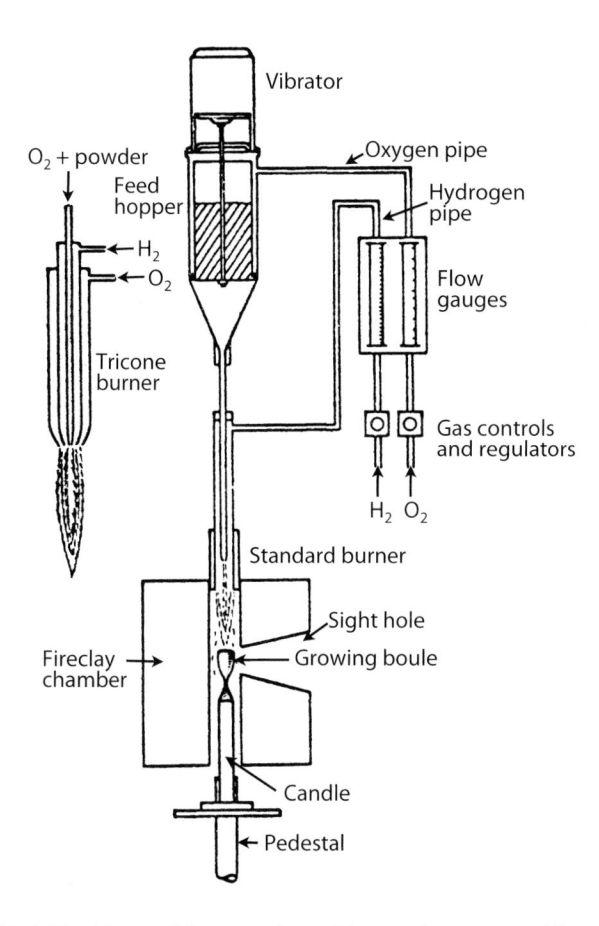

Figure 12.1 The Verneuil furnace. Inset: Tricone burner modification for titanium synthetics.

The crystal that grows on the support is known as a *boule*. It is a single crystal that is rounded, curved at the top and straight sided, but the lower third tapers inward. Under magnification the top is seen to consist of a mass of tiny crystals. When the desired growth has been achieved, the gases are shut off instantly. The actual corundum boule is one single crystal individual, but it is highly strained and is allowed to split in two pieces vertically from top to bottom when the stalk is nipped off with pliers. If this spontaneous splitting does not occur, it is given a slight tap.

Table 12. 2 Impurity-caused colours in synthetic corundum

Colour and variety	Impurity
Colourless – white sapphire	None
Red – ruby; pink – pink sapphire	Chromium
Blue – sapphire	Titanium and iron
Yellow – yellow sapphire	Nickel
Orange – orange sapphire, also known as 'padparadscha'	Nickel, chromium and iron
Green – green sapphire	Cobalt, vanadium and nickel
Yellow-green – yellow-green sapphire	Nickel, iron and titanium
Purple – purple sapphire	Chromium, titanium and iron
Violet – violet sapphire	Chromium, titanium and iron
Colour change – alexandrite simulant	Vanadium

The synthetic gems cut from these boules have all the physical and optical properties of a natural stone, including the same SG, RI, hardness and dichroism. They are made in all the colours in which natural corundums are found. If only pure alumina is used, colourless synthetic sapphire is made. The red of ruby is obtained by the addition of 1–3% chromium oxide. Blue sapphire is made by the addition of titanium and iron. Other impurities are used for different colours. Synthetic corundums with asterism, star ruby and star sapphire are produced in the usual manner, but with the addition of 0.1–0.3% titanium oxide, TiO_2. The titanium oxide is soluble at the production temperature and the boule is cooled in the usual rapid manner. The boule is then heated for twenty-four hours at 1300°C, a temperature at which the titanium oxide is no longer soluble and it precipitates out in the form of fine rutile needles arranged in three sets at 120° to one another, parallel to the first order prism faces, and when the stone is cut as a cabochon with the optic axis perpendicular to the base of the cab a six-rayed star stone is produced. Synthetic star stones are usually fairly easy to recognize with experience, as the rays are often too perfect. In natural star stones the rays are typically diffuse and may be uneven or crooked. The sharp, intense rays of the synthetic are different. Furthermore, in natural star stones the star appears to be slightly below the surface whereas synthetic stars appear right on the surface.

Despite the fact that synthetic corundum has almost every character of the natural gem, there are differences that allow discrimination between the two. With synthetic ruby, one of the first observations may be that the colour is not quite right. This is due to the fact that boules develop strain and are split into two to relieve it. As a result, the table is usually cut parallel to the vertical axis, whereas natural stones are cut with the table 90° to the vertical axis. Thus, because synthetics are cut in the wrong direction they tend to show the secondary colour of the dichroism: slightly orangy. Another general observation is that they are comparatively too clean, as natural rubies typically have inclusions. Synthetics also lack the oriented silk of fine natural rubies.

Verneuil synthetic corundums show curved structure lines due to the discontinuous feed of the powder and the rounded shape of the top of the boule. In synthetic ruby these are fine lines that look like the grooves on a vinyl record album. In synthetic sapphire these are wide and more diffuse. These structure lines, termed *curved striae*, are seen when viewed sideways-on and are best seen when the stone is immersed. A natural stone almost invariably shows some signs of its natural origin, such as straight lines or bands crossing one another at an angle of 120° or characteristic inclusions that are primarily mineral – zircon, rutile, spinel, mica, etc. – and liquid formations. Synthetic corundum often shows included gas bubbles, which may be rounded or 'flask' or 'tadpole' shape. This is diagnostic evidence of synthetic origin. Very tiny bubbles like black dots may appear as clouds that sometimes follow the direction of the curved bands of colour. The synthetic corundums of today are comparatively clean, and the large gas bubbles seen in the earlier types are absent, although curved striae are generally visible if care and patience is taken with the examination.

The synthetic ruby spectrum is the same as the natural ruby spectrum and is not useful for discrimination. In natural-blue sapphires the spectrum is diagnostic with three iron bands: 450nm, 460nm and 471nm. With synthetic-blue sapphires, the 450nm band does not normally show, or it may appear significantly thinner and weaker than the band in the natural spectrum. Natural and synthetic rubies glow crimson colour to UV light, but synthetics may sometimes be detected by their stronger fluorescent response. With synthetic sapphires, a green or sometimes blue glow under SWUV is helpful as natural blue sapphires are inert due to iron content, although heat-treated sapphire may show zoned fluorescence.

Verneuil flame fusion – synthetic spinel

After Verneuil's initial success with the production of synthetic red corundum, an attempt was made to make blue sapphire. It was thought at that time that cobalt was responsible for the blue colour of sapphire, hence cobalt was used with alumina in the first experiments. The resultant boule was found to be patchy in colour and scarcely simulated the colour of natural sapphire. Based on a 1908 report by Frémy's former student L. Paris, stating that blue sapphire was produced by adding cobalt with magnesium oxide to the feed powder, Verneuil grew a number of synthetic crystals by this method. The result was a good clean stone of uniform colour, but on examination it was found to be a spinel and not corundum. At that time there was no apparent necessity for the production of synthetic spinel; with the exception of the red variety, spinel did not become commercially significant until the 1930s. It is now an important synthetic, although it does not simulate itself; synthetic spinel is made in colours that more closely approximate other species: sapphire, emerald and topaz to name a few. Synthetic spinel is made in a colourless form and a variety of colours: colourless is pure; red, brown and green are produced with the addition of chromium; pink with copper; yellow, brown and red with manganese; blue with cobalt; blue, green, brown and pink with iron; and differing shades with other combinations. The resulting colour depends on the impurity, its concentration, and whether the atmosphere is oxidizing or reducing. There is also a colourless type in which a fine precipitate of the excess alumina forms on annealing, producing a sheen to simulate moonstone. Such stones show a bright blue-white fluorescence under SWUV.

The boule resulting from Verneuil production of synthetic spinel shows flat four-sided forms, an outward expression of its cubic crystal structure. It may also show cleavage cracks. Another key difference is that natural spinel has equimolecular proportions of magnesia and alumina; its formula is $MgO \cdot Al_2O_3$. In synthesis, it was found that a 1:1 ratio yielded boules that spontaneously fractured, so synthetic spinel is made with 1.5 to 3.5 parts alumina to 1 part magnesia; its general formula is $MgO \cdot 2Al_2O_3$. This difference is manifest in slightly higher constants: RI of 1.728 for the synthetic as against 1.718 for the natural, and SG of 3.63 as opposed to 3.60. The excess alumina strains the crystal lattice and this is observed with the polariscope as an anomalous extinction of stripes. This is *tabby extinction*, as named by Anderson.

Synthetic spinel rarely shows curved colour bands and is remarkably free from bubbles. Included solid crystals of alumina may be seen under magnification as well as strain knots and worm-like gaseous tubes. Synthetic spinels are rarely made in colours reminiscent of the natural varieties and would more likely be mistaken for another stone altogether. Higher RI and SG must arouse suspicion. Blue spinels made to imitate blue zircon and aquamarine are coloured with cobalt and appear red or orange through the Chelsea filter. Natural blue zircon and aquamarine appear green. Blue synthetic spinels show the typical cobalt absorption spectrum: three broad bands centred on 630nm in the orange-red, 580nm in the yellow, and 540nm in the green. No natural mineral shows this absorption spectrum.

Synthetic red spinels differ from other synthetic spinels in that they are equimolecular without excess alumina. The SG approximates 3.60. RI ranges from 1.722 to 1.725 and these higher values are due to excess chromium. Reds usually show pronounced curved colour lines and many gas bubbles. The fluorescence spectrum has one line dominant at 685nm, which differs from the 'organ pipe' spectrum characteristic of natural red spinels.

Modified Verneuil flame fusion – synthetic rutile

Synthetic rutile was first produced in 1948 on a modified type of Verneuil blowpipe. Synthetic rutile, TiO_2, is a tetragonal mineral synthetically produced in a number of colours including red, orange, yellow, blue, and blue-green. Despite being a diamond simulant, synthetic rutile is never colourless and characteristically shows a yellowish tinge resulting from strong absorption in the deep violet at 430nm. Titanium oxide, TiO_2, loses so much oxygen in synthesis that the boule will not grow unless extra oxygen is added during the fusion process. In order to overcome this, the Verneuil apparatus is fitted with a *tricone* burner that allows an outer envelope of oxygen to surround the flame so that two tubes carry oxygen and one tube carries hydrogen. Even so, the resulting boule is black and must be annealed at 1000°C in an oxygen-rich environment to become clear and transparent. It is also synthesized by the Czochralski pulling technique.

Synthetic rutile's trademark feature is its enormous dispersion, 0.285, seven times greater than diamond, and the stones show so much fire it makes them easy to recognize with experience. It is tetragonal and therefore anisotropic, with an RI of 2.62–2.90, and an extremely high birefringence of 0.287. The effect can be seen as a doubling of the facets so strong that it is obvious to the naked eye, and it can be identified on this feature alone. It has a hardness of about 6 and an average SG of 4.25. It is inert to both LWUV and SWUV. Synthetic rutile is quite brittle and typically shows the standard signs of wear. It is usually flawless, but gas bubbles may be present in some stones. It reached its peak production in 1955 and was soon surpassed by strontium titanate as the diamond simulant of choice.

Modified Verneuil flame fusion – strontium titanate

Strontium titanate, $SrTiO_3$, originally produced in 1953, was the first man-made gem to have no known natural counterpart at the time of its synthesis. Later, in 1987, grains of the natural mineral – known as *tausonite* – were found in Russia, but since the natural is not gemmologically significant, the term 'synthetic' is not usually applied to strontium titanate because it is not necessary to distinguish the synthetic from the natural. Strontium titanate is a diamond simulant manufactured by the Verneuil flame fusion technique using an inverted oxy-hydrogen blowpipe. Like synthetic rutile, the titanium loses oxygen and the tricone burner must be used; the boule is black and must be annealed at 1000°C in an oxygen-rich environment to become clear and transparent. Commercial production began in the United States in 1955 and it was widely circulated in the trade during the 1960s and 1970s. Since the introduction of cubic zirconia to the marketplace, it is no longer produced in quantity.

Strontium titanate crystallizes in the cubic system as a colourless, transparent material. Its value as a diamond simulant is based on its vitreous to sub-adamantine lustre, and the fact it is isotropic and can be produced in a completely colourless variety. It has only moderate hardness, 5.5 to 6.0, a relatively heavy SG of 5.13, an extremely high dispersion of 0.190, and an RI of 2.41, almost exactly that of diamond. As with synthetic rutile, the dispersion, in this case over four times that of diamond, is easily identified as unnatural with the naked eye. Strontium titanate is relatively soft and this is observed in features such as rounded facet edges, prominent polish marks, and other indications of poor wearability. Gas bubbles from the flame fusion process or features resembling ladders or centipedes may be present also due to the flame fusion process. It is inert to UV light. Strontium titanate is convincing in smaller sizes where the excessive fire is not as obvious. Its most useful application is in combination with another material as a doublet; very convincing diamond simulants are composed of a strontium titanate pavilion with a crown of harder material such as synthetic sapphire or synthetic spinel, which also serve to partly mask the excessive fire.

Czochralski pulling

Several synthetics are produced by *Czochalski pulling*, properly described as *pulling from the melt*, including synthetic corundum, the garnet-structured synthetics and synthetic alexandrite. The technique is able to yield very high quality under strictly controlled conditions. Most of these crystals are used in optical and laser industries. The starting material is melted in an iridium crucible. Platinum cannot be used as it melts at 1774°C and corundum melts at 2050°C, whereas iridium has a high melting point of 2442°C. A small seed crystal of corundum, a few millimetres in length, is touched to the surface of the melt. Strict temperature control is mandatory; it must be kept just above the melting point. If it is too high, it is capable of melting the seed crystal; and if it is too low, it is possible the crucible contents will solidify. The seed crystal is rotated about thirty times per minute as it is slowly vertically withdrawn from the melt. Solidification on the seed begins and a rod-like crystal is formed. If the crystal has a low melting point the technique can be performed in the open; if it is one of high melting point, such as corundum, the crucible must be surrounded with stabilized zirconia in powder and ceramic forms to prevent excessive heat loss. Sapphire can be grown at the rate of 6–25mm an hour. For identification, there may be very fine striations, and as there are essentially no inclusions, the absence of natural inclusions in a stone provides information.

Czochralski pulling – garnet-structured synthetics

Garnet-structured simulants are not silicates like natural garnets, but oxides with a similar structural arrangement of atoms: $X_3Y_5O_{12}$. They were originally synthesized for laser and technical purposes, but due to their transparency and lustre they became diamond simulants – the most popular diamond simulants of the 1960s. Garnet-structured simulants are manufactured by the Czochralski pulling technique as well as flux melt technique. There are a series of such 'garnets' incorporating various rare-earth elements, but the most important are the yttrium aluminium garnet (YAG) and the gadolinium gallium garnet (GGG). Both are isometric, as is garnet. YAG and GGG are colourless when pure.

Yttrium aluminium garnet, $Y_3Al_5O_{12}$, has a hardness of 8.5, an RI of 1.834, a dispersion of 0.028, and an average SG of 4.55. It is often flawless but when inclusions are present, they are typically gas bubbles or thread or tube-like features. The particular types of inclusions found depend on growth method. If grown by the Czochralski pulling technique, the common inclusions are angular black crystals and drop-like inclusions. If grown by the flux melt method, the common inclusions are flux droplets, feathers or crystals. YAG is inert to weak orange under LWUV and SWUV, but a reliably bright mauve to X-ray stimulation. YAG shows a striking rare earth spectrum. YAG is doped to produce coloured diamond simulants: chromium produces green – which is also a very good demantoid garnet simulant; cobalt produces blue; titanium produces yellow; and manganese produces red. Green-coloured YAG may appear red through the Chelsea filter and a chromium spectrum may also be seen.

Gadolinium gallium garnet, $Gd_3Ga_5O_{12}$, was never as popular a simulant because it was much more expensive to grow and is not hard enough to stand up to daily wear. GGG has a maximum hardness of only 6.5, an RI of 1.97, an extremely high SG of 7.05, and a dispersion of 0.045, which is almost exactly that of diamond. The body colour is often slightly brownish. GGG exhibits a weak peach colour to LWUV, a strong strawberry colour to SWUV, and a lilac glow to X-rays. Characteristic inclusions are: triangular or hexagonal metallic platelet inclusions of iridium contributed from the container during crystal pulling, and gas bubbles. Specimens show typical signs of wear, including abraded facet edges and scratches.

Skull melt – cubic zirconia

Cubic zirconia, CZ, is the best and most common diamond simulant encountered in the trade. It has attained this status through its many advantages: optical properties close to diamond resulting in a fire and brilliance nearer to that of diamond than any other simulant; it is inexpensive – priced on average at less than US$1 per carat depending on size and colour; available in quantity; and can be produced in most cuts, sizes, shapes and colours.

Table 12.3 Cubic zirconia colour range by dopants

Dopant	Colour
Chromium	Olive
Cobalt	Lilac
Iron	Yellow
Vanadium or thulium	Green
Erbium or europium or holmium	Pink
Praseodymium	Amber
Cerium	Yellow, orange, red
Nickel or titanium	Yellow-brown

Cubic zirconia is zirconium oxide, ZrO_2. In nature, zirconium oxide forms a monoclinic mineral variety called *baddeleyite*, found in Sri Lanka. Only the monoclinic structure is stable at room temperature and pressure, and although it is proper to refer to the man-made material as synthetic cubic zirconia, the term 'synthetic' is seldom applied because the natural variety is not stable at room temperature.

The commercial production process for cubic zirconia was developed in 1972 and stones entered the market in 1976. A unique manufacturing process, known as a *self-contained melt* or a *cold melt*, had to be developed to accommodate the high melting point of pure zirconium oxide, 2750°C. A temperature this extreme precluded the use of other crucibles because even nickel and platinum melt at lower temperatures. *Skull melt* technology was developed in the USSR. Colourless material is available as well as coloured material produced by doping with allochromatic transition metals. Colour is affected by both the concentration of dopant and its oxidization state; therefore, combinations of dopants can be used to produce specific shades that make very good coloured diamond simulants or coloured stone simulants.

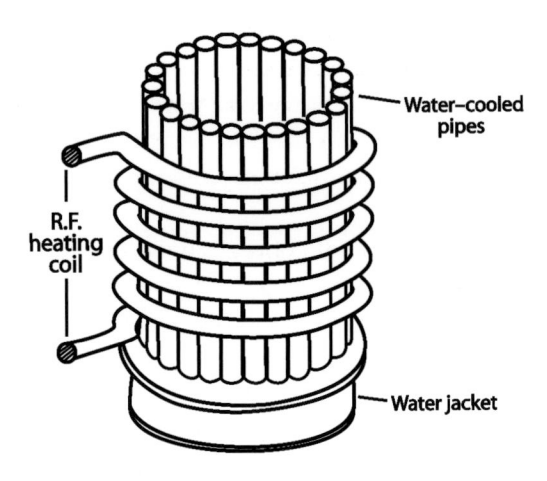

Figure 12.2 Skull melt apparatus.

The basic skull melt apparatus is an open-topped cylindrical cup comprising vertical hollow copper tubes for walls, employing a circulating water system to cool the walls during crystal growth and contain the melt. A water-cooled copper coil heater, activated by a radio-frequency generator, surrounds the walls. There are small gaps between the vertical copper tubes to allow energy to pass through into the cup. The cup is filled with zirconia powder mixed with a stabilizer, either CaO or Y_2O_3, to prevent the material from reverting to monoclinic form on cooling. Pieces of zirconium metal are added to the cup because ZrO_2 is an electrical insulator and will not heat without the presence of metal. The zirconium metal is heated first and this heats the adjacent zirconia powder, which begins to conduct electricity, quickly spreading heat. The zirconium metal itself melts, reacts with oxygen in the surrounding air, and produces more zirconium oxide. As the material in the cup is heated to 2800°C and melted, the material at the outside, that in contact with the cool copper tubes, remains unmelted and forms a shell – the self-contained melt; the molten zirconia is contained, in effect, within a shell of its own powder. As the melt progresses, a layer like a porous crust forms at the top of the melt. This reduces heat loss, as heat must be maintained for several hours to allow impurities to vaporize. Once the melt has remained molten for the required amount of time, the cup is lowered out of the heating coil and slow cooling begins. Cubic zirconia crystals nucleate on the bottom of the cup and grow upwards until the entire melt solidifies, producing a mass of crystal columns that can be separated by tapping. The crystals grow highly strained, and annealing at 1400°C in an oxidizing atmosphere removes residual strain. A typical skull of 30 cm in diameter can yield in excess of 50kg, or 250 000ct. Individual crystals can be produced large enough to yield faceted stones of several hundred carats.

The chemical composition of cubic zirconia is typically ZrO_2 with 10 mol. percent Y_2O_3 or 15 mol. percent CaO. It crystallizes in the isometric system. The physical and optical properties of cubic zirconia vary slightly due to the type and proportion of stabilizer. The SG of Y_2O_3 stabilized CZ averages 5.95 and the SG of CaO stabilized CZ averages 5.65. The RI has a mean value of 2.16. Dispersion is 0.060 and hardness is 8–8.5. Specimens of finer quality are free from inclusions, but on occasion small particles, gas bubbles and a striated appearance may be seen with magnification. With experience, cubic zirconia is easy to recognize. It differs from the diamonds it simulates in several ways: the girdle appears different from the sugary adamantine lustre of a diamond girdle; facet junctions are commonly slightly rounded; rare earth spectra have been reported; and small particles and sometimes cracks are present. With unmounted stones, the high SG is obvious and the lustre is vitreous rather than the adamantine of diamond. UV response is often very weak under LWUV and a faint yellow under SWUV.

Solution techniques – synthetic emerald

Beryl gems have the basic composition $Be_3Al_2Si_6O_{18}$. The colouring agent is chromium oxide: Cr_2O_3. Vanadium produces a similar colour and many natural and most synthetics contain some vanadium. Synthetic emerald is grown by flux techniques: flux transport, flux reaction and flux melt. Flux is a melting solvent. Materials of high melting points can be dissolved in a flux of a much lower melting point, such as lithium molybdate with a melting point of 705°C, even if that flux is at room temperature.

Flux transport

Gilson emeralds are grown from a water-free salt bath at atmospheric pressure. The crucible is composed of two halves: one for crystal growth and one which contains the molten salt solvent that replaces solvent lost by evaporation during slow growth. The beryl dissolves in one section of the crucible, in the hotter region at right, and deposits on the emerald seeds in the cooler region at left. Mechanical stirrers aid the flux convection currents. A small temperature gradient is used to keep supersaturation low, which prevents spontaneous nucleation. The flux is lithium molybdate-vanadate. Growth is initiated on a natural colourless beryl seed which becomes coated with synthetic emerald top and bottom and then the seed is cut away and the synthetic emerald growth is used as the seed in the actual growth runs. Seed crystals are approximately 4cm x 1mm and growth rates average 1mm per month.

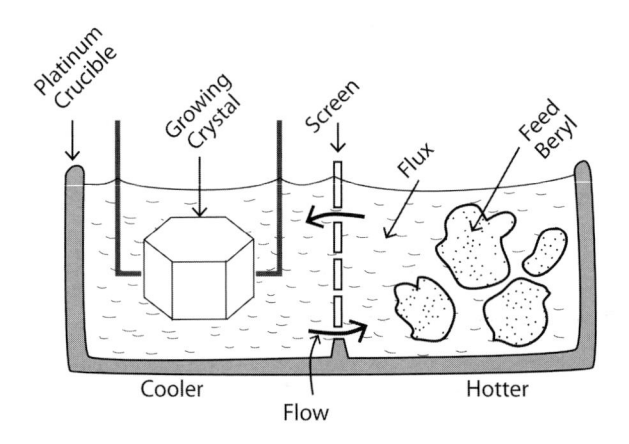

Figure 12.3 Synthetic emerald growth by flux transport.

Flux reaction

With the flux reaction technique, growth proceeds by diffusion. Chatham synthetic emeralds are speculated to be grown by this process in a lithium molybdate-vanadate flux. Slabs of silica glass, SiO_2, float on the surface and as the flux is heated they dissolve in the flux at the top and slowly diffuse downwards. Beryllium oxide and aluminium oxide, plus chromic oxide as a colouring agent, are at the bottom, and when the flux is heated they melt and move upwards through the flux by convection heat current. When they meet in the centre they react to form emerald in solution. As the beryl solution becomes supersaturated, emerald precipitates out and crystallizes on the seed crystals. It is mandatory that the SG of the flux be carefully controlled to remain lighter than the bottom ingredients and denser than the growing emerald seeds and the silica at the top. A platinum shield prevents the emeralds from floating up to the silica. Crystals up to 2cm in length can be grown in twelve months.

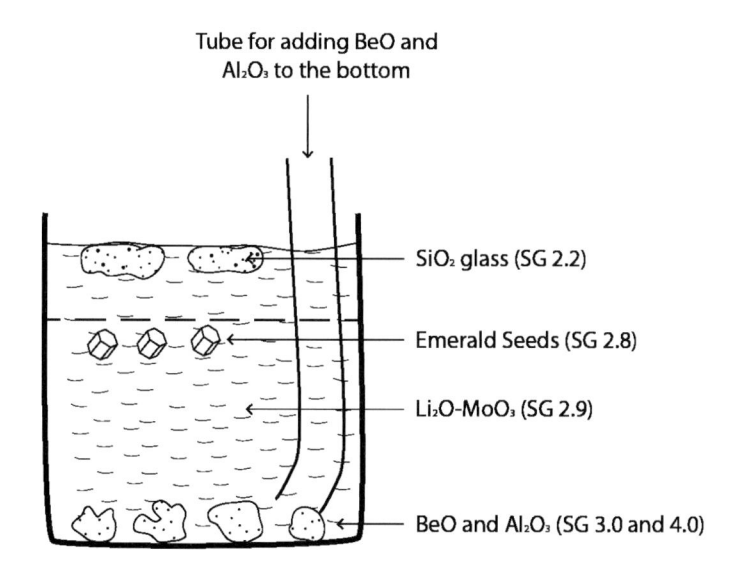

Figure 12.4 Synthetic emerald growth by flux reaction.

Flux melt

The flux melt technique allows growth by spontaneous nucleation of crystals. Flux melt technique is often confused with flux reaction, but *no seed crystals* are used in flux melt and the entire container is gradually cooled. This technique, although termed 'flux melt', is not to be confused with the general category of melt techniques such as Verneuil flame fusion; this is categorically a *solution* technique. Some emeralds as well as ruby, spinel, quartz, alexandrite, YAG and GGG can be produced by this method. To produce ruby, alumina plus chromic oxide for colouring are dissolved in a flux at 1350°C in a closed platinum crucible. The melt is cooled at the rate of 4°C per hour and in a matter of days the ruby begins to crystallize out. It takes about eight days for the entire contents to crystallize, at which point the molten flux may be poured off.

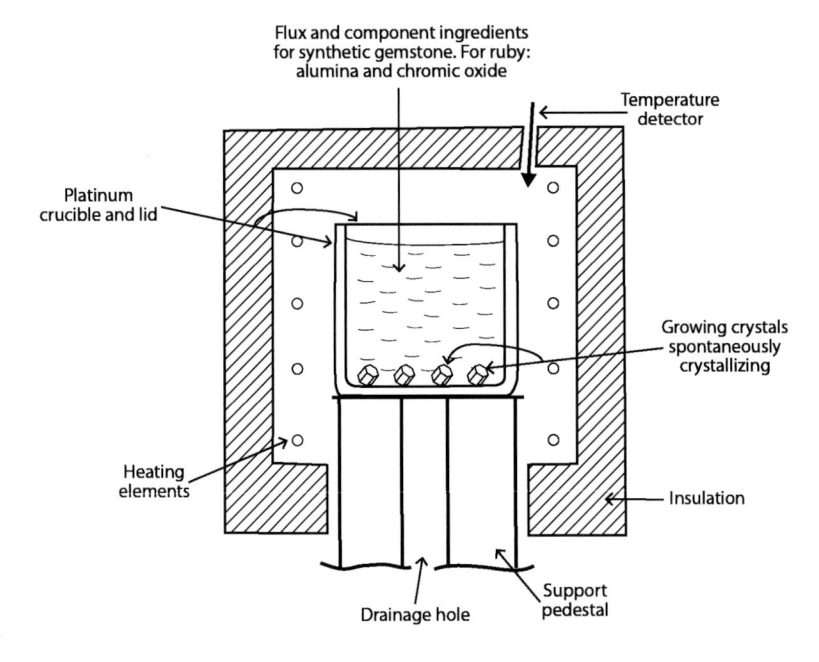

Figure 12.5 Synthetic ruby growth by flux melt.

Flux melt rubies may show 'fingerprint' inclusions and larger flux inclusions. Some rubies appear oversaturated with colour and this is often not very natural looking. The colour of synthetic emeralds often tends to be a highly saturated slightly blue-green, often showing intense red through the Chelsea filter. This resembles the finest of Colombian stones, but not the appearance of most emeralds in general. The physical constants of flux synthetics are typically slightly lower than naturals: synthetic SG averages 2.65 and most naturals have an SG above 2.70, synthetic RI averages 1.561–1.564, with a birefringence of 0.003 to 0.004, and the lowest natural is usually 1.570 upwards with a birefringence of 0.005 or 0.006. Phenakite inclusions are diagnostic. Veil or lace-like feathers, usually curved, are often seen. The inclusions are the most diagnostic features. They are usually filled with residual flux, but may also contain gas.

Hydrothermal growth

Hydrothermal growth techniques use water under high pressure to dissolve the constituents for the synthetic gemstone. Water under high pressure is able to readily dissolve minerals such as beryl and quartz, thus both emerald and quartz are produced by these methods: emerald by hydrothermal reaction and quartz by hydrothermal transport. Lechleitner of Austria produced the first hydrothermal emerald in 1960. By hydrothermal means he grew a surface of synthetic emerald on an already cut pale-coloured natural beryl. The stones were lightly polished afterwards. The colour is not intense and the coating is quite thin and is easily seen when the stone is immersed. Owing to the constants of the beryl used, the density may vary. The density, usually about 2.69, that of pale aquamarine, is lower than most emeralds at 2.70, but it may be higher and in the case of some colourless and pale-pink beryls it may then reach to over 2.80. A series of parallel fissures creating a crazing pattern may be seen in the surface layer. The inclusions are those of beryl, but not those of emerald.

Hydrothermal reaction – emerald

Hydrothermal reaction growth is by diffusion-reaction assisted by convection. In this way it closely relates to flux reaction growth and similarities can be seen from their diagrams. The hydrothermal method depends upon the fact that water at a high pressure will boil at a much greater temperature than 100°C. At high pressures water can have much higher temperatures and be capable of dissolving substances, such as the silicates. The temperature is raised at the base to 360°C and the water boils under great pressure. To produce emerald, beryllium oxide and aluminium oxide are dissolved at the bottom and silica is dissolved at the top. Convection occurs as well as diffusion and they meet in the middle and produce a supersaturated solution, so that because it is in a cooler region of the autoclave, the mineral constituent will come out of the solution and form on the seed crystal. The silica-containing nutrient must be kept separate from the other ingredients to prevent unwanted nucleation and to restrict growth to the surface of the seed crystals.

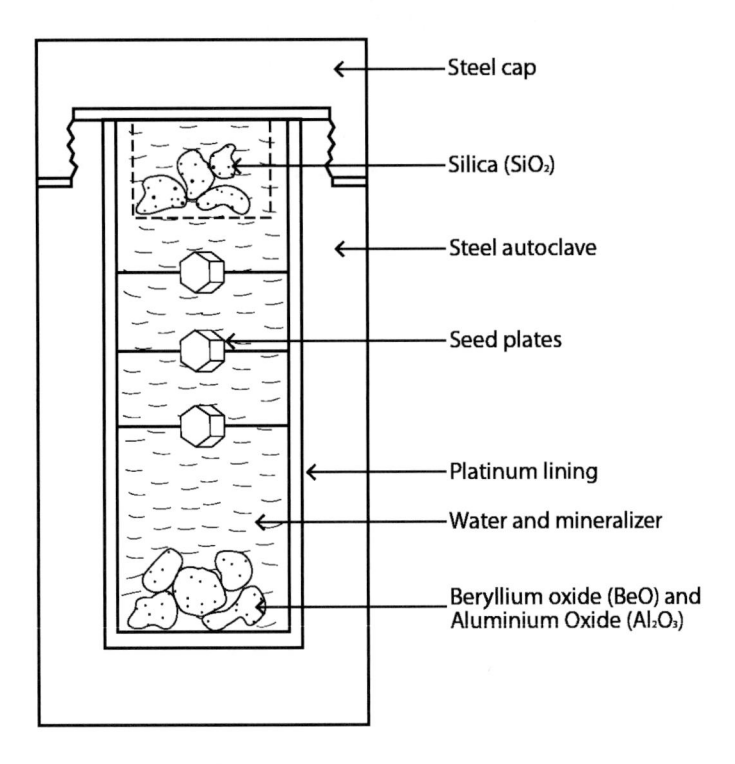

Figure 12.6 Synthetic emerald growth by hydrothermal reaction.

The colour is often too saturated as it is with the flux synthetics and the stones also show intense red through the Chelsea filter. Hydrothermal emeralds are the hardest to identify, as their constants are much closer to natural emerald. The SG is slightly lower, averaging 2.68 to 2.70, and RI averages 1.566–1.580, with a birefringence of 0.005 to 0.006. They may contain spiky nail-head inclusions containing liquid and gas capped at the broad end with phenakite crystals. All hydrothermal stones and all natural emeralds have a very strong absorption in the deep infra-red region of the spectrum. This is not the case with flux synthetics.

Hydrothermal transport – synthetic quartz

Synthetic quartz is grown by hydrothermal transport. The method uses a steel pressure vessel, termed an *autoclave* or *bomb*, with a silver lining. It is filled 85% full with water. Quartz is grown from a hot water solution containing sodium silicate with crushed natural quartz as the feed material. The bomb is tightly sealed and heated. This causes the water to

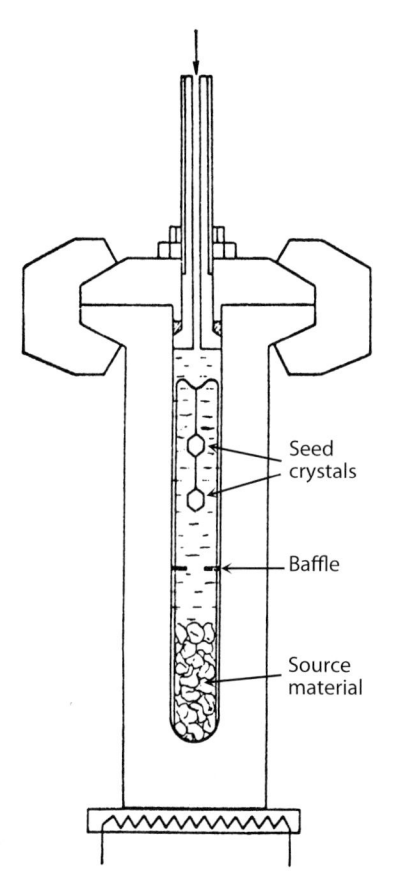

Figure 12.7 Synthetic quartz growth by hydrothermal transport.

expand and increases the pressure, which in turn increases the boiling point. At 200°C all the water becomes steam and the pressure rises to 1400 bars as the steam tries to expand. The temperature is raised to 400°C and a mineralizer, such as sodium hydroxide or sodium carbonate, must be added to achieve solubility good enough for crystal growth. Crushed quartz is placed at the bottom and thin seed plates are supported on a silver framework in the upper region of the bomb. The bomb has a temperature gradient of 40°C between the hotter feed section and the cooler seed crystals. A saturated solution of quartz in the mineralizer forms at the bottom and it rises as convection current in the upper, cooler region. Here the fluid cannot hold all the quartz in solution, causing the excess to crystallize on the seed crystals. Maximum growth rates are 1mm per day. Blue can be grown by adding cobalt; yellow and green by adding iron; and amethyst is produced by adding iron followed by irradiation.

Synthetic quartz is hard to identify. If colour zoning is seen, it is a sign it is probably natural. The seed plate may be visible, which is positive proof. Synthetic amethyst may contain 'breadcrumb' inclusions. Natural amethyst is twinned and the synthetic is grown untwinned and polariscopic examination of the polysynthetic twinning may be observed.

High-pressure solution – synthetic diamond

It was first thought that diamond, being a single element, pure carbon, should be easy to synthesize, but early experimentalists suffered from incomplete knowledge until diamond was found associated with kimberlite rocks, the weathered necks of ancient volcanoes, proving that the dense atomic arrangement of diamond forms deep within the earth and that a combination of extreme temperature and pressure was essential.

The first step was to understand the nature of diamond. Sir Isaac Newton compared diamond's dispersion to that of oil and concluded that if oil is combustible, then diamond must also be combustible. In 1673, Robert Boyle discovered that diamond vaporized when subjected to high temperatures. In 1772, Antoine-Laurent Lavoisier focused sunlight on to a diamond sealed in a glass jar filled with oxygen and found the sole by-product to be carbon dioxide – the same by-product produced by burning charcoal. In 1797, Smithson Tennant burned identical weights of diamond and charcoal and converted them to the exact same volume of carbon dioxide gas, thereby establishing chemical equivalence. Scientists were slow to accept that carbon, the same element that formed soft black graphite, could form diamond. Ultimately, diamond was accepted to be pure carbon and, since its specific gravity is higher than that of graphite, it was speculated that pressure, because it decreases volume and increases specific gravity when it acts on a substance, might convert other forms of carbon to diamond.

In 1880, Scottish chemist James Ballantyne Hannay conducted his famous experiments using a mixture of 1% lithium, 10% nitrogen-rich bone oil and 89% hydrocarbon in the form of paraffin. The mixture was placed in a wrought-iron tube, welded shut at the ends, and placed in a massive furnace with concrete walls. It was fired at red heat for several hours, breaking down the paraffin into carbon and hydrogen; the hydrogen caused an enormous, and highly dangerous, pressure build-up. On most occasions, the tubes leaked or exploded. Several times his entire furnace was blown apart. His experiments failed to produce diamonds.

F. F. H. Moissan began studying diamond formation in nature in 1889 and found that in each diamond-bearing material he studied, iron was common; and he concluded that iron, or a metal like it, was the key to diamond synthesis. Since iron expanded as it solidified, he thought he could use this force to produce the high pressure required for converting carbon to diamond. He created the Moissan electric-arc furnace, consisting of a pair of carbon rods inserted into a groove between two blocks of lime with a central cavity for the sample. His technique was to heat iron in the form of an iron cylinder 2cm in length and 1cm in diameter, surrounded with sugar charcoal in a carbon crucible. When it was heated to the 3000°C capacity of the furnace, Moissan believed the heat would flow from the iron to the carbon and transform the carbon to a volatile carbon phase from which it could be converted to diamond when the mixture was cooled; the iron contracted from the outside inwards, inducing high internal pressure on the solution of carbon in the liquid core. He tried air-cooling, plunging the white-hot crucible into cold water, and also quenching the iron into a bath of molten lead hundreds of degrees cooler than the crucible. This last method produced an iron lump, which, when dissolved in acid, revealed tiny diamond-like crystals. He died believing he had succeeded, but this was later disproved. In recent years researchers have repeated his experiments to determine what he did produce and it turned out to be silicon carbide, a present-day diamond simulant that has been named moissanite.

In 1906, Sir William Crookes combined Hannay's sealed steel tube method with the brief temperatures of 5000°C and pressures of 8000 atmospheres achieved by Sir Andrew Noble when heating steel tubes packed with gunpowder or cordite until they exploded. Crookes declared the combined technique produced tiny crystals resembling diamond, but we now know his conditions were insufficient for diamond production. Crookes had most likely produced silicon carbide as well.

When the carbon phase diagram was produced, first by Rossini and Jessup in 1938 and later modified by others, it showed that at room temperature and pressure up to 10 000 atmospheres, graphite is the stable polymorph of carbon. At the higher temperature needed to break atomic bonds, making carbon atoms mobile in order to change structure and produce diamond, the required pressure has to be much higher and the conditions must remain in the stability field for diamond: the high pressure must be maintained while the temperature is lowered in order to avoid reversion to graphite.

Nobel Prize-winning American physicist Percy Bridgman developed a new design that would change high-pressure research for ever: a specialized unsupported area seal and the Bridgman anvil. With his new inventions he eventually extended the pressures achieved to more than 400 000 atmospheres or 6 million lb/in^2. He tried to make diamond by squeezing graphite under increasing pressure between two strong surfaces. None of his experiments formed diamond, but he did achieve conditions approaching those necessary for synthesis.

General Electric (GE) took up the task because they needed industrial diamonds to form their carbide products. They codenamed their effort 'Project Superpressure' and made it a top priority. On 16 December 1954, an experimental run by H. Tracy Hall succeeded. The key was to reach the temperature and pressure *at the same time* as the metal was in a liquid state. The GE process for commercial industrial diamond, grit, employs a temperature of 1400°C–1800°C and approximately 55 000 atmospheres, a pressure equivalent to 800 000 lb/in^2. The reactant is a layered disc of ingredients – graphite, iron, manganese and vanadium – arranged in a graphite tube with metal end plates, top and bottom. The reactant is packed in a pyrophyllite cylinder dusted with a thin layer of hematite to further increase friction. Pure pyrophyllite is plastic under extreme heat and flows to fill in gaps, acting as an insulator and a seal between the anvils as the pressure is applied. Pyrophyllite can be lathed to shape and at very high pressure its melting point rises from 1360°C to 2720°C. The reactant is placed in a high-pressure device, which is subjected to extreme pressure using a hydraulic press. The insulator properties of pyrophyllite allow an electric current to be passed from the upper anvil, through the metal end plates and the graphite, to the lower anvil. The nickel end plates partly melt and increasing amounts of graphite are dissolved into solution. After the carbon solution is saturated in the graphite stability field, the temperature and pressure are further increased, and the carbon becomes supersaturated in the diamond stability field. The reaction is a type of slow convection, which moves the carbon flux across a thin film of nickel skin. Diamond is precipitated at the ends. Total conversion takes just a few minutes. Synthetic diamond grit is produced at the rate of 0.33mm per second.

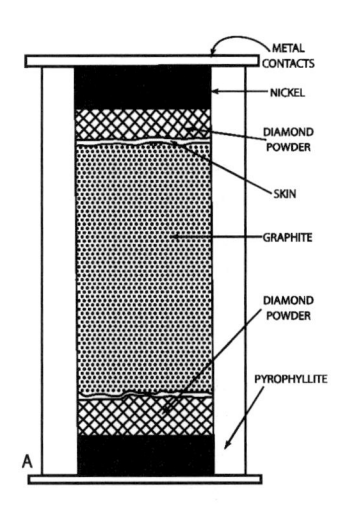

Figure 12.8 The GE reaction chamber for the production of synthetic diamond grit.

In grit production, diamond grows as carbon dissolves from a sample of graphite and a molten metal. The exact concentration of carbon is not strictly controlled, so numerous crystals grow at the same time. A different process is used in the production of gem diamond mono-crystals of a size and quality suitable for cutting. They are generally grown for specific high-tech uses. Essentially, a different reaction chamber is used and three changes are made: (1) diamond is the feedstock; (2) seed crystals are present in the form of thin slices of natural or synthetic diamond; and (3) there is an induced temperature gradient of 30°C between the middle and ends of the chamber. In summary, the temperature gradient results in higher solubility of carbon in the molten centre section. The carbon in solution travels by convection to the ends. The molten metal catalyst surrounding the growing crystal moves towards the hotter centre of the cell, and now, in saturation, the carbon precipitates out by crystallizing layer by layer on the seed. To produce gem-quality monocrystals, growth must proceed slowly over a period of days or weeks. Generally, a 6mm cube or octahedron can be produced in sixty hours. Because diamond grit is used as the feedstock, the process is sometimes referred to as the *reconstitution method*. Note that since the resulting diamond is synthetic, it would be inappropriate to call it reconstituted diamond without also being qualified as synthetic.

Properties such as dispersion, RI, SG and hardness are of no diagnostic value. The first step in differentiation of synthetic diamond should always be a close examination of the inclusions. Transparent crystalline inclusions and coloured mineral inclusions are never found in synthetics and their presence is proof positive of natural origin. Synthetic diamonds exhibit metallic inclusions of flux metal. If the metallic lustre of the flux is seen, one can be assured of synthetic origin. During synthetic manufacture of a coloured diamond the distribution of impurities is uneven, causing characteristic colour zoning appearing as square, octagon, cross-shaped, columnar or funnel shaped. The distribution of colour in natural diamonds is predominantly even. Synthetic diamond graining is a phenomenon exclusive to coloured stones. In natural near-colourless diamonds the graining may sometimes be coloured and found along the planes parallel to the octahedral faces forming a cross-hatched or 'tatami' pattern. In coloured synthetic diamonds, the graining reveals cubo-octahedral growth as hourglass, Maltese cross or stop sign formation.

Natural diamonds typically exhibit a stronger response to LWUV than they do to SWUV. Synthetics characteristically show a relatively stronger response to SWUV; however, both synthetic and natural-blue diamonds have a stronger SWUV response. Near-colourless synthetics are always inert to LWUV. The distribution of fluorescence is usually even in stones of natural origin and uneven in stones of synthetic origin. Generally, phosphorescence is present to some degree in a synthetic diamond – often persistent, lasting 30–60 seconds. Cathodoluminescence is representative of internal growth structure: natural diamonds characteristically show an even distribution, a product of their slow and even layered growth, and synthetics tend to exhibit uneven distribution, showing the angular arrangement of internal growth sectors. Most often some sectors are inert.

Natural colourless and near-colourless diamonds, Type Ia stones, exhibit diagnostic absorption bands: the most pronounced at 415nm, and others at 423nm, 435nm, 452nm, 465nm and 478nm. This Cape Series spectrum is diagnostic for natural diamond, as is the 415nm band regardless of whether other bands are visible. Synthetic near-colourless diamonds tend to display no significant spectra, typically an increasing general absorption at the violet end of the spectrum.

Ferromagnetic metals such as iron, cobalt and nickel are the most favoured metal catalysts for synthetic production and even if a synthetic diamond has no visible magnetic inclusions, it will still contain residuals from the metal catalyst that causes it to be attracted to a magnet. The absence of magnetism is not proof-positive of a natural diamond, but the presence of magnetism is proof-positive of synthetic origin. Only natural Type IIb diamonds are able to conduct electricity whereas synthetic colourless, near-colourless and blue diamonds are all semi-conductors. There are also specialized instruments available to positively differentiate natural and synthetic diamonds. These instruments, the DiamondSure™, which detects the presence of the 415nm band, and DiamondView™, which analyses internal growth structures, are best used in combination and are suitable for testing parcels, as well as individual mounted and loose diamonds.

In addition to high-pressure synthesis, diamond for experimental and technical uses can be produced by low-pressure chemical vapour deposition, CVD, which has great commercial potential in thermal management, cutting tools, optics, electronic devices and wear-resistant coatings.

High-pressure sublimation – synthetic moissanite

Synthetic moissanite is silicon carbide, SiC, a compound popularly known in its polycrystalline form as carborundum, an abrasive. Moissanite, named after synthetic diamond experimentalist F. F. H.

Moissan, occurs naturally on earth and in meteorites, but not in adequate size to facet. Synthetic moissanite is a relatively new diamond simulant; worldwide distribution began in mid-1998. It remains expensive relative to other simulants, in the order of several hundred dollars per carat. Charles & Colvard Ltd, formerly C3 Inc., of the United States, holds the patent protected production process for synthetic moissanite. The company estimates annual production to be 250 000ct.

Moissanite is produced by a high-pressure seeded sublimation technique in which the SiC is vaporized and crystallized directly to a solid without passing through a liquid state. The synthesis takes place in a radio-frequency-heated graphite furnace in which a porous graphite cylinder is inserted into a graphite crucible. The uniform grain size feed of SiC is placed between the walls of the cylinder and crucible. The seed crystal is placed slightly off-axis at the bottom of the crucible in the coolest region. As the SiC is vaporized, it passes through the graphite cylinder and crystallizes on the seed. Initiated in a vacuum, the process is changed to a low pressure argon atmosphere as growth proceeds. It occurs with a feed temperature of 2300°C and the seed crystal kept 100°C cooler.

Synthetic moissanite is hexagonal. Hardness is 9.25, and although this is still significantly less hard than diamond, 10, it is by far the hardest of all simulants. Moissanite wears well, even better than sapphire. It has high reflectance and high thermal conductivity, and these present the real cause for concern because it can be confused with diamond on traditional and less sensitive thermal and reflectivity testers. Dispersion is 0.104, SG is 3.22, and RI is 2.648–2.691, with a birefringence of 0.043.

In reality, cubic zirconia is a better diamond simulant – it looks more like diamond than synthetic moissanite does; however, the introduction of moissanite to the marketplace caused widespread panic because it was able to fool those who relied on testers to identify diamond from its simulants. With experience, synthetic moissanite can be identified by a number of features. It is anisotropic with a strong birefringence, and doubling is evident when the culet is viewed through any bezel facet. Note that doubling of the culet is not seen through the table because the stone is commonly cut with the optic axis perpendicular to the table. As a general observation, the body colour always tends to be a bit greenish. The spectrum shows absorption from the deep violet at 425nm into the ultra-violet. Inclusions appearing as fine hair-like needles are characteristic of moissanite. These open, empty tube channels, or micropipes, are hexagonal in cross-section and parallel or sub-parallel to one another, elongated in the direction of the optic axis. More recent production shows fewer tubes, but their presence remains diagnostic. Some moissanites show scattered pinpoint inclusions that may form clouds. Synthetic moissanite is inert to LWUV or SWUV.

LESSON 13

Simulants

The last lesson discussed synthetic gems, manufactured products that have natural counterparts with essentially the same chemical composition, crystal structure and properties. The subject of this lesson is simulants – those stones that have only the superficial appearance of the stones they imitate, but not their chemical composition, crystal structure and properties. Simulants may be natural, treated or man-made materials. For example, ruby simulants may be natural gems such as garnet; treated gems such as diffusion-treated pale corundum; synthetic gems such as synthetic red spinel; man-made material such as glass; or an assembled stone such as garnet-topped doublet. It is important to note that by this definition synthetic ruby is not classed as a ruby simulant. The terms 'simulant' and 'imitation' may be used interchangeably, but in recent years favour has been with the term *simulant*, because *imitation* has taken on negative connotations. Information on natural gems, treated gems, misrepresented species, and synthetic gems is available in their respective sections and will not be discussed here. The subjects of this section are man-made materials, such as glass and plastics, which are used to simulate gemstones. Identification of glass is covered in this lesson; details on the identification of particular simulants made of other materials are covered in the discussions of the gemstone species. Assembled stones are covered in the next lesson.

Glass

Unlike synthetic gems, which are a product of the scientific ingenuity of the nineteenth and twentieth centuries, glass – except for the few natural glass gems – is the historical simulant. It has been used in jewellery to simulate various stones since ancient times. It is worth noting that it was also used in jewellery as a gemstone in its own right many centuries ago. Glass was used extensively to simulate diamond, although it is rarely used for this purpose now as many other inexpensive simulants are marketed that are closer in appearance and much harder; however, it is often seen imitating coloured stones. The term *paste* is used indiscriminately to refer to all glass imitations of gems.

By standard definition, glass is an inorganic amorphous solid material. Technically, it is a *supercooled liquid*. That is to say, it is a liquid, but it is below the freezing point. Glass is a vitreous solid formed by rapid melt quenching, meaning that the cooling is rapid – in relation to the cooling time for crystalline solids – preventing crystallization, and instead a disordered atomic configuration is frozen into the solid state below the glass transition temperature (although some of the constituents, even after considerable time, may crystallize in a process known as devitrification). So in fact, at ambient temperature and pressure, glass is in a metastable state, but as we know it, it is rigid in our everyday experience, causing us for all practical purposes to consider it a solid.

The basis of most glasses is silica, SiO_2, which is always prone to forming glassy substances on cooling. There are two main types of glass: (1) *crown glass* or *calcium glass* – such as window or bottle glass; this is a mixture of silica, potash, soda and lime. These are the most common glasses. Pure silica can be made into glass but the other additives are to aid production. It has a range of refractive index from 1.44 to 1.54 and a specific gravity ranging from 2.05 to 2.60. Although it is fairly hard for a glass, Mohs 5.5 to 6, it has little lustre or dispersion; and (2) *flint glass* or *lead glass* – consisting of silica, potash, soda and lead oxide. The greater the lead content the higher the RI, dispersion and SG; however, the hardness of the glass decreases in proportion to lead content. It will be recalled that this is the type of glass used on refractometers. Lead crystal is glass with a minimum lead content of 24%. Lead glasses have a range of RI from 1.50 to 1.80+ and the SG ranges from 2.56 to 6.00+. The lead content also increases the liability to tarnish due to atmospheric influences that cause a chemical alteration in the glass. The figures given reflect the full ranges of the two types of glass; but when used as imitation gemstones, the ranges are actually much narrower, typically: crown glass RI 1.52 to 1.54 and SG 2.53 to 2.57; and flint glass RI 1.58 to 1.68 and SG 3.15 to 4.15.

In general, all glass is amorphous and isotropic with a refractive index between 1.47 and 1.70. Its RI, SG, hardness and other properties vary with its constituents, most notably lead content. Dispersion, also variable with lead content, may be as low as 0.01 or as high as 0.09, but it averages 0.04. With higher RI and higher dispersion glasses, the hardness averages 5.

Specialized glasses are also produced. Beryl glass, known as fused emerald, has the emerald composition produced by melting the constituents and cooling the melt. The result is a material softer than emerald and with lower constants than emerald: an RI of 1.52 and an average SG of 2.46. Other beryl colours are also produced: cobalt produces an aquamarine simulant and didymium produces a pink morganite simulant. Similarly, fused quartz has an RI of 1.46 and an SG of 2.21, in contrast to natural quartz that has an RI of 1.54–1.55 and an SG of 2.65. Fused beryl and fused quartz have lower constants than their natural counterparts because they are non-crystalline. They are also typically harder than other glasses, averaging Mohs 7. Borosilicate crown glass, in which some silica is replaced by boric acid, is characterized by an RI of 1.47 to 1.51 and an average SG of 2.35. Pale blue produces an aquamarine imitation. Calcium-iron glasses have constants similar to beryl: an RI ranging from 1.57 to 1.59 and an SG ranging from 2.66 to 2.75. Table 13.1 gives a summary of these glass families and their constants.

Table 13.1 Glass by type

Glass type	RI	SG
Opal	1.44 to 1.46	2.07 to 2.15
Titanium-iron	1.47 to 1.49	2.40 to 2.52
Borosilicate crown	1.47 to 1.51	2.30 to 2.37
Beryl	1.515 to 1.516	2.44 to 2.49
Calcium crown	1.52 to 1.54	2.53 to 2.57
Calcium-iron	1.57 to 1.59	2.66 to 2.75
Flint/lead	1.58 to 1.68	3.15 to 4.15

The colour of a glass imitation gemstone is usually produced by various metal oxides: cobalt oxide produces blue, chromium oxide produces green, copper or selenium oxide produces red, manganese oxide

SIMULANTS

produces violet, and purple of Cassium, a gold compound, produces a ruby red colour. When boron or aluminium oxides are added it gives the glass greater resistance to heat or chemical attack. Some of the more important colouring compounds or elements, along with the colours they produce, is presented in Table 13.2. Combinations of two colourings are possible and the final colour achieved depends on: (a) the colouring agent; (b) the atmosphere – whether oxidizing or reducing; and (c) annealing after the glass has cooled. Translucent to opaque glasses used to simulate stones such as turquoise are produced with the addition of an opacifier. Frequently, glass stones are set with a backing of metallic foil. The foil reflects light and creates far greater brilliance than glass itself could achieve.

Table 13.2 Glass colorants

Colour	*Colourant*
Violet, mauve, purple	Manganese or nickel in potash glass
Blue	Cobalt
Blue-green	Copper or iron in the reducing condition
Green	Chromium, iron in the ferrous state
Yellow-green	Chromium compounds; uranium compounds; iron in reducing condition
Yellow	Silver salts; titanium compounds; cadmium sulfide; sulphur; combination of iron and manganese
Brown-yellow	Sulphur with carbon; iron oxide; uranium in quantity
Browns	Iron; nickel when in soda glass
Pinks	Selenium under certain conditions
Ruby Red	Copper with controlled reducing conditions
Red (with orange tint)	Selenium or selenium compounds
Red (with purplish tint)	Gold
Red	Iron oxides in the ferric state
Smoke tints	Platinum or iridium

Colour does not appear to have any strict relationship to the RI value in general; however, it is rare to find red glass with an RI below 1.62. Commonly, reds, pinks and some shades of green show the highest figures. Table 13.3 offers the combined values of a series of typical glass imitations.

Table 13.3 Glass by colour, RI, SG and type

Colour	RI	SG	Type of Glass
Colourless	1.47	2.30	Borosilicate crown
Yellow	1.498	2.43	Calcium crown
Pale blue	1.50	2.36	Borosilicate crown
Pale blue	1.51	2.37	Borosilicate crown
Pale blue	1.51	2.46	Calcium crown
Blue	1.515	2.44	Fused beryl
Emerald green	1.516	2.49	Fused beryl
Colourless	1.54	2.87	Light flint (lead)
Pale blue	1.575	3.19	Flint (lead)
Pale blue	1.59	2.70	Calcium-iron
Ruby red	1.63	3.69	Flint (lead)
Yellow	1.633	3.627	Flint (lead)
Colourless	1.635	3.74	Flint (lead)
Dark purple	1.64	3.76	Flint (lead)
Sapphire blue	1.645	3.80	Flint (lead)
Orange-brown	1.68	4.12	Flint (lead)
Pink	1.68	4.07	Flint (lead)
Red	1.683	4.16	Flint (lead)
Emerald green	1.70	4.25	Flint (lead)
Yellow	1.77	4.98	Flint (lead)

Beginners may feel at a loss identifying glass, as it is one of the instances in which the constants are seldom relied upon. SG is rarely used as a test in general, and with glass it is almost never necessary. RI is useful to some extent. Glass is isotropic and therefore non-birefringent. Any stones that show a single shadow edge reading between 1.50 and 1.70 on the refractometer should be suspected of being glass. Any such stone has a far greater likelihood of being glass than anything else. All other natural gemstones with readings between 1.50 and 1.70 are birefringent, such as beryl and topaz, and therefore will show measurably different RI values on two shadow edges. The exception is jet, RI 1.66, and it is expected that jet can be identified by other means. Always be certain that RI is properly tested in monochromatic light, whereupon it will be obvious whether the stone is birefringent or not. Knowing how to take a proper refractive index measurement will eliminate all such confusion. Pastes are non-birefringent, and hence also monochromatic, although weak dichroism is a trait of some natural stones in that RI range so this result is of little use conclusively. However, if the stone shows two colours with the dichroscope it cannot be glass.

With experience, identification of a glass is usually straightforward with a loupe or microscope magnification. All glasses used as gemstone simulants are characterized by a low degree of hardness. A test given in every early textbook was that paste imitations could be scratched with a steel file and most genuine gemstones would not be affected. There are many non-destructive ways to identify glass and this is a destructive test that should not be performed. Relative hardness can be examined with the loupe. Signs of poor wearability are usually obvious. These include small nicks and chips, scratches and a general dulling of the surface. The ordinary quartz particles in dust cause rapid abrasion of the surface.

Bubbles are present in almost every specimen. Typically they are round, but they can also be oval or elongated – a pointed ellipsoid is very characteristic. They can appear individually, in groups, chains or sheets, and vary in size from tiny to relatively large. Another characteristic is the presence of swirl lines caused by the flow of molten material and incomplete mixing of constituents. These swirl lines, termed *curved striae*, are caused by streams of material of differing RIs. Most glass imitations are moulded, and this produces rounded facet edges that are easily identified with the loupe. Moulding also results in concave facets, resulting from the glass cooling and the surfaces caving inwards. Unpolished girdles often show what is termed an 'orange peel' effect because the surface tends to pucker as it cools. Fissures and cracks may also appear, and small star-like markings may be seen which are due to incipient devitrification of the glass. Polishing marks are usually prominent.

Glass has vitreous lustre and this is seen most visibly on conchoidal fractures, although many of the gems it imitates have similar lustre and fracture. Glass is a poor conductor of heat and feels warm to the touch or on the tongue in comparison with the natural gemstones it imitates; this is a most pronounced distinction from diamond as diamonds are excellent heat conductors. Most pastes glow under SWUV and not under LWUV. The strong apple-green fluorescence seen with many paste stones, generally yellow-green colours, is usually attributed to uranium although manganese in certain states also produces this rather spectacular fluorescence. The fluorescence is similar to that of yellow-green synthetic spinel.

Table 13.4 Identification of glass

Constants
1. The RI and SG are not correct for the simulated stone
2. Non-birefringent and never shows two shadow edges on the refractometer – a single RI in the 1.50 to 1.70 range is a strong indicator
3. Relatively soft: hardness 5 to 6

Surface observations
4. Shows wear such as scratches and abrasions
5. Colour is often not right for the simulated stone
6. Evidence of moulding: rounded edges and concave facets
7. Conchoidal fractures
8. Fissures, cracks and star-like markings due to incipient devitrification may be evident

Internal features
9. Bubbles often present: small to large, rounded to elongated, individual to multiple
10. Curved striae due to incomplete mixing
11. Random striae due to rapid cooling

Other
12. Never shows pleochroism
13. Maltese Cross on the polariscope – showing strain
14. Comparatively warm to the touch
15. Low reading on thermal conductivity meter
16. May exhibit fluorescence under SWUV and not under LWUV

Cobalt-coloured blue glass has a diagnostic spectrum with three broad bands: one centred on 655nm in the red, one on 590nm in the yellow, and 535nm in the green, with the centre band as the narrowest of the three. They are rich red under the Chelsea filter. Other blue glasses exhibit no distinctive spectrum and show dirty green under the Chelsea filter. Selenium-coloured red glasses show a spectrum with a single broad band in the green. Those pink and red glasses coloured by rare earth elements may show the characteristic rare earth spectrum of fine bands.

The term 'goldstone', a double misnomer because it is neither 'gold' nor a 'stone', is a glass simulant of sunstone that has bright spangles due to small triangular or hexagonal platelets of copper crystals. The copper is dissolved in the molten glass and, upon cooling, is reduced and separates out in the form of minute metallic crystals. Usually brown in colour, it can be made with inclusions other than copper to produce alternative colours, such as a blue variety, or with copper inclusions in a different coloured base glass.

A cat's eye imitation is made with thin glass optical fibres, stacked in cubic or hexagonal arrays, in an amorphous matrix. This imitation, known as *Cathay stone* or *catseyte*, comes in various colours and shows a sharp, bright, straight chatoyant streak. This streak is far too perfect and gives away the simulant. The bundles of fibres are usually seen with the loupe in a direction perpendicular to the chatoyant streak.

When fluorides and phosphates are added to a glass containing lime it produces a moonstone simulant. The lime is the key ingredient as it causes the calcium compounds to precipitate to yield the desired translucency.

Pressing white opaque glass into a mould with a star-ray formation embossed on it simulates star stones. The cabochon, with the star rays impressed into it, is coated with a thinly applied deep-blue glaze that causes the star to appear below the surface. Alternatives are to etch the rays on the back of the cabochon or to cement rayed foil to the back of the glass cabochon.

Opal is simulated by a material known as *opal essence*, formerly known as Slocum Stone, a glass produced by a controlled precipitation process. It has been said that the vivid colours are produced from the tinsel-like laminated material that diffracts light from the spacing between the laminates; however, the tinsel-like laminate is shredded in appearance and not regularly spaced, as one would expect for diffraction. It is most likely that the mechanism that yields this colour phenomenon is the same mechanism that yields the colour-flash that is diagnostic of filled cracks in diamond and emerald: 'dispersion staining', where the thin layer has a similar RI to the host material but a significantly different dispersion. Opal essence has an RI of 1.49 to 1.50 and an SG of 2.4 to 2.5. If the interference colour is not too gaudy, it can be convincing. Since it is a glass, the characteristic glass bubbles and swirl marks remain evident.

Other materials

Another material popular for simulants is the group of artificial resins popularly known as the *plastics*. They have an extremely low degree of hardness, averaging between 1.5 and 3, and a very low SG averaging from 1.05 to 1.55. The RI ranges from 1.5 to 1.6. Most are made by the injection moulding process and most are sectile. Plastics give off a sharp characteristic odour when touched with a hot point. It is common for plastics to imitate amber and opal, and the details for those simulants and their identification are given in their respective sections.

Some semi-opaque gems and ornamental stones may be imitated in porcelain, a medium that is particularly suitable for imitations of turquoise. Porcelain is baked uniformly fine clay. It is usually moulded and glazed. Its SG lies in the range 2.1 to 2.5, with an average of 2.3. Very few ornamental gem materials are around this value; only sodalite and thomsonite could be confused with it.

Ceramics are polycrystalline solids produced from finely ground inorganic powders that are heated, fired or sintered, and occasionally compressed. Sometimes a low melting point binding agent is added to help the particles adhere to one another. As with porcelain, it is common for the surface to be glazed. Gilson manufactures a ceramic lapis simulant and turquoise simulant.

Hematite is sometimes simulated by metals, most notably a steel-grey titanium dioxide with hardness of 5.5 and SG approximately 4.0. It can be identified by its yellow-brown streak, as the streak of hematite is red. Also, fine-powdered lead sulfide with added silver, with a hardness of 2.5 to 3 and an SG of 6.5 to 7, also makes an effective simulant.

LESSON 14

Composite Stones

Composite stones, alternatively called assembled stones, have been known since Roman times. Composite stones are assembled fabrications of two or three pieces of material typically held together with a layer of cement such as epoxy resin. Composite stones are generally described as being either: (a) *doublets* – when the stone consists of two main pieces; or (b) *triplets* – when the stone consists of three pieces of material joined together. By far, the greater majority are doublets. There are some differences in the interpretation of these terms as *soudé-type* stones are called doublets in Europe and triplets in North America. The separation plane may be in the crown, at the girdle, or in the pavilion. Composite stones have been used for centuries and are almost always encountered mounted.

There are several reasons for the production of composite stones: (1) to produce a better wearing and more lustrous surface on a glass simulant by using a crown of harder material; (2) to produce an apparently larger stone from a given genuine material; (3) to produce a stone of better colour and appearance; (4) to utilize material that would be otherwise unsuitable for jewellery such as with opal mosaic triplet; and (5) to provide a support base or protective crown for thin and fragile gem materials. This is the case with opal as it is often found only in thin seams and the doublet/triplet provides support for the thin slice of opal and allows it to be used in jewellery. It is often said that composite stones are not always fraudulent and meant to deceive, and this can be the case with opal composites, although it is the disclosure of the composite and not the composite itself that determines whether there is fraudulent intent. In any situation where the composite is not disclosed and an attempt is made to pass off the stone as genuine material of one piece, it is fraudulent.

True doublets, sometimes called *genuine doublets*, consist of two parts of the same colour and same species of genuine stone cemented together. They can be made of diamond on diamond, emerald on emerald, ruby on ruby, and sapphire on sapphire – although the only one that is commonly encountered is opal on opal where a thin slice of colourful precious opal is cemented to an underlying piece of *potch* (opal not showing a play of colour).

Semi-genuine doublets are made with a crown of the material being imitated attached to a pavilion of an alternative material. The most important stone of this type is the *diamond doublet* in which the crown of the stone is a piece of real diamond and the pavilion is another colourless material – typically, glass, quartz, topaz, colourless synthetic spinel, or colourless synthetic sapphire. If unset, immersion in a heavy liquid such as di-iodomethane will reveal the difference in relief between the diamond crown and pavilion material as a result of the differences in RIs. A caution with immersion methods: heavy liquids can erode the cement layer, causing visible pseudo-flaws or even separation of the two components. Such tests should be conducted quickly and the stone cleaned afterwards. These doublets are usually bezel set so that the setting covers the join of the material. Holding the doublet with the table facet tilted slightly away from the observer, a dark border to the opposite edge of the table facet will be seen owing to the reflection of the facet edge on the cement layer. In some cases, prismatic colours are seen in the cement layer. This is caused by air films that have penetrated the cement layer at the areas of deterioration.

Opal doublets may be genuine or semi-genuine depending on the backing material. Typically they consist of a thin slice of precious opal with good play of colour cemented to a backing of common opal, or if the effect of a black opal is desired, a backing of black onyx or black glass. Black onyx or black glass is easily identified as it looks very different from the back of a black opal, but white doublets may be more difficult to detect. Many natural opals are cut with the natural backing of the original seam wall and if this is fairly straight it can resemble a doublet. When such a stone is set it is usually necessary to unset it to be certain. Looking through the back, most opal doublets are dark and, with magnification, flattened bubbles at the junction may be seen.

Semi-genuine doublets are rare with the exceptions of diamond and opal doublets; however, others do appear in the marketplace such as a crown of natural greenish-yellow sapphire backed with either synthetic

COMPOSITE STONES

Figure 14.1 Illustrations of composite stones.

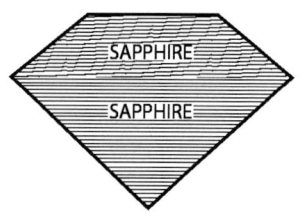

A true doublet with genuine sapphire on genuine sapphire

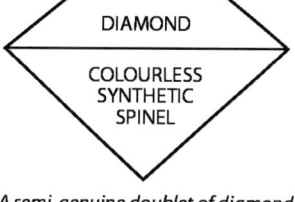

A semi-genuine doublet of diamond on colourless synthetic spinel

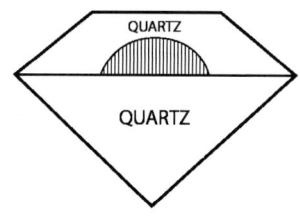

A hollow quartz doublet with a coloured material in the centre

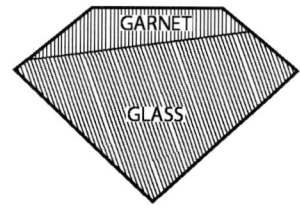

A garnet-topped doublet with almandine garnet on coloured glass

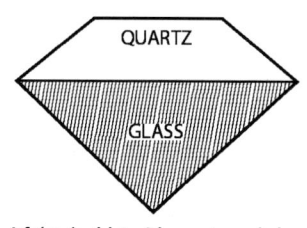

A false doublet with quartz and glass

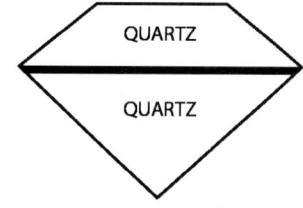

A soudé stone in which the crown and pavilion are both colourless quartz and the centre is a coloured layer, typically green to imitate emerald

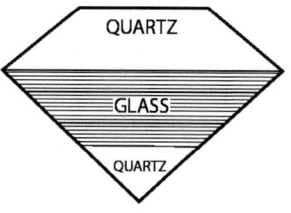

A triplet with quartz for the crown and pavilion tip and a large section of coloured glass in the middle

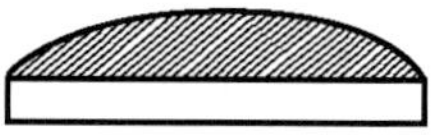

An opal doublet with a thin top section of precious opal attached to a base of common opal

An opal triplet, which is the opal doublet with a cap of quartz

blue sapphire or with synthetic ruby. Care is required not to be hasty in judgement, as natural zoning and silk in the crown, as well as the 450nm absorption band, would seem to indicate natural sapphire. If unset, observation of the join is the best detection. Curved bands and gas bubbles of synthetic corundum may be visible in the pavilion. And under UV light the synthetic sapphire base glows greenish and the synthetic ruby base glows crimson. In both cases, the crown is inert.

False doublet refers to a composite stone where the material being imitated is entirely absent. Often they consist of a quartz crown or other colourless material cemented to a base of coloured glass. These versions are often sapphire blue, ruby red, emerald green, purple or yellow. The danger here is that if the glass is citrine or amethyst coloured, the crown of quartz will give a refractive index reading for the genuine stones. If it is possible to take a reading from the pavilion it will be an isotropic reading typical of glass. Under UV light the cement layer often fluoresces strongly. And if tested, the SG of the stone will not be correct for quartz; however, it is more likely that upon immersion to test SG, the nature of the composite is revealed by observation in the liquid.

Hollow doublets are composite stones in which a cavity is hollowed out of the piece forming the crown and this is filled with suitably coloured liquid and then closed in on the back with a pavilion of the same material as the crown. This is usually quartz and, in general, these composites are rarely found. They can be detected easily as the liquid-filled cavity can be seen when the stone is held sideways.

Imitation doublets consist of two pieces of colourless glass cemented together with coloured cement. Immersion readily reveals the coloured layer at the centre.

Garnet-topped doublets were very common, but are now seen far less often. They are produced by fusing a thin slice of almandine garnet to a piece of coloured glass and then faceting the whole material. Garnet is the only stone that will readily fuse to glass. The garnet is usually so thin in relation to the depth of the stone as to have no great effect on the overall colour of the composite, which is due to the glass. Such stones are made in all colours including colourless. They have been used to simulate sapphire, ruby, emerald, amethyst, topaz and many others. When the composite is faceted there is no particular effort made to orient the join in the girdle plane and, as a result, a characteristic of these doublets is that the plane of the join is often partway up the crown and usually at an angle to the girdle.

There are several features useful for identification. Magnification will reveal the join, as the mount cannot hide it when it is above the girdle plane. A lustre difference between the garnet and the glass is evident. If it is loose and laid table-down on a white surface, a red ring is seen around the table. This effect is of course masked if the overall colour of the stone is red. Bubbles in the join plane are often seen; these are the result of the fusing of the two materials. An RI reading for almandine garnet, 1.78, is obtainable on the crown and if it is possible to take a reading from the pavilion it will be the anisotropic one of glass. Immersion reveals a difference in relief between the two materials. Additionally, the glass may glow whitish to greenish under SWUV and this may be an obvious distinction as garnet is inert.

Soudé-type stones have been produced in several varieties. The earliest of these types consisted of two pieces of quartz, often selected with natural inclusions, which were cemented together by a layer of green-coloured gelatine. These stones were first fabricated to imitate emerald and are referred to as *soudé emerald*, but to be technically correct they are in fact quartz doublets. They suffered from the fact that the green cement deteriorated to yellow with age and one was left with an imitation citrine instead of an imitation emerald. Owing to the type of dyestuff used in the layer, these stones showed a red residual colour through the Chelsea filter. Immersion in water clearly reveals the colourless crown and pavilion with a plane of colour at the girdle. The RI reading from the crown is that of quartz, and under UV light the cement layer fluoresces.

Later a better modification of this soudé stone was made; again with two pieces of quartz but with a layer, as far as is known, of sintered glass of suitable colour. In this case the colour is more likely due to metal oxide and not organic dyestuff. These stones are usually green to imitate emerald. Unlike the earlier gelatine type they show green through the Chelsea filter and the coloured layer does not fluoresce to UV light.

A third type of this stone consists not of quartz top and bottom, but of two pieces of colourless synthetic spinel with a coloured layer between of sintered heavy lead glass. *Synthetic spinel doublets* are made in all colours including black. The RI reading is that of synthetic spinel, 1.73. Immersion in oil reveals a colourless stone with a layer of colour at the girdle. Under SWUV the synthetic spinel glows a strong bluish-white in clear distinction to the inert cement layer.

Recently another variation of the soudé emerald has been produced using colourless beryl or pale aquamarine in place of the quartz or colourless synthetic spinel for the crown and base. There is a thick layer of green glass or green cement between the two parts. Immersion in water will usually reveal the colour plane at the girdle. The spectroscope shows lines in the red due to the dyestuff. Another version is also made that uses genuine emerald for the top and bottom. The inclusions will be those of natural emerald but none will continue through the middle of the stone. A difference in inclusions may be a subtle clue. In this case the spectroscope may show the chromium spectrum of emerald, but also in combination with lines in the red due to the dyestuff.

Table 14.1 Detecting composite stones

Junction layer	With magnification, the junction may be visible from the side
Red ring	Garnet-topped doublets placed table-down on a white background will show a red ring around the table
Colour differences	Colour differences may be seen either side of the junction plane in stones viewed from the side
Lustre differences	Lustre differences between the two materials may be seen either side of the junction plane
Immersion	Immersion may reveal colour differences either side of the junction; or with triplets it will reveal a colourless body with the colour layer in the middle
Bubbles	Magnification may reveal bubbles at the junction plane
UV light	UV light reactions may be different for different parts of the composite; or for the coloured layer of soudé stones
Spectrum	Unusual spectrums such as those with lines in the red due to dyestuff
Refractive index	Readings for almandine garnet, synthetic spinel or synthetic corundum may arouse suspicion; and, if possible, a different refractive index for the pavilion material

Synthetic doublets consist of a crown of colourless synthetic spinel or colourless synthetic sapphire and a pavilion of one of the colourless high RI diamond simulants, such as strontium titanate or synthetic rutile. In this case the crown provides hardness and a partial masking of the excessive fire from the pavilion. To limit the fire further, it is common for the junction to be further down the pavilion. These diamond imitations will have the synthetic spinel or synthetic corundum RI from the crown and, if spinel, the SWUV light will produce a strong bluish-white colour. In reality they look significantly different to diamond and they have far too excessive a fire to be either synthetic spinel or synthetic corundum.

Triplets are constructed of three pieces, such as a crown of quartz and a pavilion tip of quartz with a thick centre section of coloured glass between them. They are composite stones that are rarely met in practice with the exception of the *triplet opal*, which is a conventional opal doublet with a dome-shaped covering of transparent quartz on the top. This cap may also be colourless synthetic spinel or colourless synthetic sapphire. They are produced to create a better-wearing opal with more brilliance as the cap acts as a type of lens; however, the effect is unnatural and they are easily identified. Another variation is a triplet made of three pieces of jadeite to imitate Imperial jade. The three pieces are a hollow cabochon of fine translucent jadeite, another cabochon smaller than this and cut to fit into the back of the hollow cab, and a flattish oval piece which closes in the back. The cabochon in the middle is coloured with a jelly-like dyestuff in the same colour as Imperial jade. It is inserted into the hollow cab and the back piece is cemented on to close it in. The spectrum reveals bands in the red due to the dyestuff.

What may also be termed a doublet or triplet is a piece of star-rose quartz cemented to a blue-coloured mirror and possibly completed with a third backing piece of some other stone. The reason for the mirror is that star rose quartz displays diasterism, meaning it shows the star effect only in transmitted light. When viewed under an electric lamp, an image of the bulb is seen at the crossing of the rays of the star. The closer the light is brought to the stone, the larger the image of the bulb becomes. The colour is not quite the same as that of natural star sapphire, and its lustre is markedly less.

In general the detection of composite stones is usually simple, provided one does not rush into making a judgement and remembers to look for the signs of a composite. One should first carefully examine the stone with a hand lens to see if there is a line at the junction. In the garnet-topped doublets this is not usually along the girdle, but halfway up the side facets of the crown. The junction can be anywhere in the crown, girdle or pavilion. Immersion in water or oil usually reveals the composite, especially soudé composites – oil giving better results with higher RI stones. Under the microscope a bubble layer may be seen in the junction plane. Testing with UV light often reveals differences either side of the junction, or reveals the coloured layer. Unusual SGs and spectrums should arouse suspicion. With experience, examination with a microscope will detect any composite stone.

LESSON 15

Gemstone Altering Treatments

An interesting aspect in the study of gems is the alteration of a gemstone's appearance and colour by various means, including the ability of certain gemstones to actually change colour in response to irradiation and/or heat. Colour alterations are of two types: (1) superficial treatments; or (2) treatments that change body colour throughout. Superficial treatments such as dyeing, staining and foiling have been documented as far back as the first century AD. Eventually such treatment evolved into the use of varnishes and enamels. Older superficial treatments have now been replaced with systematic alterations by heat treatment or high-tech treatments such as irradiation and annealing, and high-pressure/high-temperature (HPHT) processing capable of altering colour throughout the stone. These produce altered colours that fill the entire stone and are more permanent. The section concludes with two advanced diamond treatments known in the trade as fracture filling and laser drilling, which improve the overall appearance of diamond clarity.

Treatments are prevalent in the gem trade for many gem varieties and are accepted as standard practice in some cases. Most citrine is heat-treated amethyst and most pure blue aquamarine is heat-treated to remove a greenish component of the colour. Gemstone treatments can produce legitimate commercial products. Legitimacy requires that the treatment is known and accepted, or is fully disclosed in all transactions where it is not an accepted practice. Non-disclosure, and/or non-detection of treatments that may be considered fraudulent, destroy consumer confidence. Therefore, it is the responsibility of trade members to maintain current knowledge of treatments and their identification.

Dyeing, staining and foiling

Materials of a polycrystalline or cryptocrystalline nature are somewhat porous and are capable of being stained or dyed to various colours. This is commonplace with chalcedony and agates, which can be dyed or stained to add colour or increase the contrast in banding. The oldest dyes were organic and, although many were very effective, the colour tended to fade over time. A common imitation of black onyx is produced by boiling or soaking chalcedony in a solution of sugar and water for a period of weeks, followed by treatment with sulphuric acid. The sugar penetrates into the pores and is deposited there during the boiling phase. The chemical formula for sugar is $C_{12}H_{22}O_{11}$. The $H_{22}O_{11}$ portion of the formula is eleven times the formula for water: H_2O. In the interaction with sulphuric acid, an acid with great affinity for water, the water is abstracted leaving C_{12}, which is carbon, and thus the black colour is produced. Some opal matrix is treated in a similar manner and is sold as *treated black opal*. Chalcedony of a poor grey colour can be stained a variety of colours using chemical salts. The colour is permanent. The green material coloured by chromium salts can be detected because natural green chalcedony, *chrysoprase*, is coloured by nickel and is green under the Chelsea filter whereas the chromium-coloured material appears pink. Also, the spectrum of the rare natural chrome-coloured green chalcedony shows only one band in the red whereas the stained green stones show three bands.

Lapis lazuli is often stained with Prussian blue, a synthetic blue pigment, to improve the look of material with many white spots. The colour is relatively permanent although it may wear off over time. Jasper is also stained to look like lapis lazuli; such material is referred to as *Swiss* or *German Lapis*. Poorly coloured turquoise has been colour enhanced by staining. The white mineral *howlite* has also been stained to imitate turquoise of a good blue colour.

Grey or white jadeite has been stained to an *Imperial jade* colour by organic dyes. The colour is very convincing but the effect is not permanent as it has a tendency to fade in strong light and, after some months, even to fade in the dark. Jadeite is also dyed mauve, and in this instance the colour tends *not* to fade. Under magnification it is usually possible to see intense small patches of colour in the granular structure of the jadeite. Under the spectroscope the dyed green shows a broad absorption in the red not found in natural-coloured green jadeite. There is also an absence of chromium lines in this part of the spectrum and it shows red under the Chelsea filter. The mauve is usually more intense in colour than natural-coloured material and it may fluoresce orange to LWUV. Organics such as ivory, coral and pearls are also commonly dyed.

Rock crystal can be heated and, while it is hot, if immersed in coloured solutions or inks that rapidly cool it, this results in cracks that draw the dye into the fissures, producing *crackle dyed quartz*. In general, under magnification it may be possible to see dye materials where they have concentrated in pores and fissures. Also, a drop of alcohol or acetone on a cotton swab touched to an obscure place may reveal the presence of a dye or stain.

In antique jewellery it was common practice to apply a metallic foil backing to either a colourless or a poorly coloured gemstone and then mount it in a closed-back setting. With cabochon stones, sometimes the foil backing was engraved with lines to produce a star-stone effect. Diamond was commonly foiled to produce a more brilliant-looking stone.

Painting a thin layer of colour on the polished facets of a gemstone can produce or mask colour. It was documented in 1568 that the back of a yellowish-coloured diamond could be painted blue, the complementary colour of yellow, to mask the yellow and improve the colour of the diamond. To produce added colour, poorly coloured rubies, sapphires, emeralds, topazes and amethysts were frequently found with painted back facets. Fancy coloured diamonds were often imitated with painting and foiling. Opal is also found with the back painted black to imitate black opal. Under magnification it may be possible to see the uneven application of the colour, or to observe scratches in the coating. With most painted stones the colour can be removed with soap and water, but if coloured lacquers have been used it requires acetone to remove the colouring.

Impregnation

Lapis beads may be dipped in paraffin wax. It penetrates the grains and, although it adds no colour, it has the same effect as wetting the stone with water; it increases the saturation of colour and gives the appearance of a smooth, lustrous polished surface. Wax is also applied to penetrate the pores of turquoise. This reduces the light scattering from a porous surface and also protects the surface from acidic perspiration when the jewellery is worn. Waxing is an accepted practice, but if a colouring agent is added to improve the apparent value of poor colour material this is considered a fraudulent practice. Wax is also commonly applied to jade carvings.

Opal and turquoise are both stabilized by plastic impregnation. Light-coloured fractured Brazilian opal can be coloured by impregnating it with a pigmented plastic under vacuum conditions to yield a variety of colours. Transillumination by diffused light usually reveals dyed and filled internal fractures in any translucent gem.

Oil and wax are also used to reduce the visibility of cracks and surface-reaching fractures in a number of gemstones including ruby, sapphire and emerald. With emerald, almost all stones are oiled to enhance their transparency and this is accepted practice. Sometimes the oil is drawn into the cracks in a vacuum process: the emerald is soaked in warm cedar oil, RI 1.57, and inserted into a vacuum chamber to draw air out of the fractures and draw the oil into them. The oil may bleed out with gentle heating and household cleansers rather easily remove it, so the stone will be affected if worn when washing. The ultrasonic cleaner is very damaging to emerald overall, but it readily damages the appearance of oiled stones by removing the oil. Even if handled delicately, over time oil has a tendency to dry out and reveal the fractures. Low-melting-point resins, such as Canada balsam, or an epoxy resin such as Opticon, are more durable and more difficult to detect. Coloured oil or resin is also used for emeralds, rubies and sapphires, a practice dating back some one thousand years. It greatly improves the apparent colour as well as transparency, but it is considered a fraudulent practice. The colour may fade and the oil still bleeds out with gentle heating. Wiping a heated surface with a white cloth may reveal the dyestuff. Immersion and magnification should reveal a concentration of coloured dye in the cracks and fissures. The surface-reaching fissures of opals are also treated with oil, as are those of amber.

Heat treatment

Many different species of gemstones, including corundums, quartz, topaz and zircon, are altered in colour when subjected to heat treatment.

Variations of heat treatments have been performed for centuries. Today, the common commercial practices result in colour changes that are generally permanent and accepted by the trade; disclosure is not required.

With sapphire, heat treatment to 1600°C in a reducing atmosphere can develop or intensify blue colour in stones with titanium as it converts ferric oxide to ferrous oxide. This is typically performed on oily-looking greyish stones. In an oxidizing atmosphere, 1600°C heat treatment can lighten or remove blue colour by converting ferrous oxide to ferric oxide. Heating for an extended period at 1300°C causes exsolution of titanium oxide with the formation of rutile needles, therefore producing or intensifying asterism. Heating to 1600–1900°C can cause the rutile needles in translucent to opaque corundum to dissolve and, if cooled quickly, within a few hours, they do not have time to exsolve and recrystallize and the stone becomes much clearer. Pale yellow or colourless material containing ferrous oxide can be turned golden yellow by extended heating to 1000–1400°C in air. Heat treatment results remain permanent. Detection of heat-treated sapphire is sometimes possible. If there is prolonged heating, it may be possible to see pockmarks on the surface; these are often most visible on the girdle when polishing is incomplete. The strength of the 450nm iron absorption band is diminished in heat-treated stones. Characteristic silk may be virtually absent and inclusions may expand, causing circular stress fractures. Some stones may exhibit chalky green to whitish fluorescence under SWUV that is often zoned in segments that coincide with colour zoning.

A temperature of 1200°C in air can lighten the colour of brownish-red Thai rubies. It can also drive off the bluish coloration from purplish rubies and pink sapphire.

Topaz is very commonly heat-treated. Heating to 500–600°C changes yellowish-brown Brazilian material to rose pink colour upon cooling; lower temperatures produce salmon pink. The colour is attractive and permanent. It can be detected by the difference in dichroism: heat-treated stones show pink and colourless, and naturals show pink and shades of orange-yellow.

Another regularly heat-treated stone is zircon. Colourless, blue and golden-yellow can be produced from reddish-brown material. The resulting colour is affected by the nature of the rough, the degree of heating, and whether the atmosphere is oxidizing or reducing. A reducing atmosphere, such as a crucible filled with wood charcoal, which gives an atmosphere of carbon monoxide, turns the stones blue. In an oxidizing atmosphere red, golden-yellow or colourless is produced. A considerable amount of these stones are by no means colour stable: sometimes the blue will fade to a brownish tint after exposure to strong sunlight. The colour can be restored if the stone is heated in air to 900°C. Heat-treated stones show a different absorption spectrum: there are weak bands at approximately 650nm, 580nm and 560nm – and also a slight general absorption at 450nm in the violet.

Much amethyst is heated to produce yellow citrine or brown or reddish-brown colours. To produce citrine the amethyst must contain ferric-iron and the temperature must be raised slowly from 150°C to 400°C. Heat-treated citrine is non-dichroic, whereas the dichroism is pronounced in naturally coloured citrine and smoky quartz. Amethyst with ferrous iron can turn green, producing prasiolite at temperatures of 400–500°C.

Green and greenish-blue beryl turns aquamarine-blue when heated at 250–500°C. At 400°C, orange and yellow-orange beryl is altered to pink morganite. Tourmaline is also heat-treated. Notably, dark blue-green stones lighten when heated to 650°C to a bright emerald-green colour tinged with blue. Tanzanite owes its fine blue colour to heat treatment of brownish-green material. And heating also drives off the undesirable yellowish tinge found with some kunzites.

Surface diffusion is a variation of heat treatment that alters the composition of the surface layers of corundums by means of a diffusion process undertaken at high temperatures. Pale or colourless corundum can be treated to simulate fine coloured ruby or sapphire by this process. The faceted stones are packed in a clay mixture containing the required transition element additives: chromic oxide for ruby, and both ferrous oxide and titanium oxide for sapphire, and heated to 1750°C for a period of days. It is also possible to induce surface asterism with titanium dioxide as an additive. The stone must be given a final polish post-treatment. It is not an accepted practice as the colouring is only skin-deep and can be removed by normal repolishing. Diffusion-treated stones lack the characteristic colour zoning typical of ruby and sapphire as colour appears fairly evenly distributed, although cases have been reported of poor-quality zoned sapphires being diffusion treated to improve their colour. Immersion in water or a heavy liquid may reveal a concentration of colour on the facet edges as the fine layer of colour is usually noticeably thinned on the faces of the facets during the final polishing. It may also be possible to see inclusions that have expanded, causing circular stress fractures surrounding them, and the 450nm band is absent.

A more recent process developed in Thailand in 2001 is the lattice diffusion of corundum in which light pink sapphires from Madagascar were heated in an oxygen atmosphere to the prized pink-orange padparadscha colour. This new process, which is evidenced to occur at higher temperatures than surface diffusion, is capable of creating colour through up to 80% of the stone. It is known that beryllium is present in these stones and is not present in those of natural colour. It is capable of other colour changes including: yellow-green to yellow, dark red to orangy-red, and light yellow to dark yellow. Stones may be identified by surface pitting, inclusions with stress fractures surrounding, and the colour zoning may reveal a core of a different colour.

Irradiation

One of the most important methods of colour change in gemstones is irradiation and irradiation followed by heat treatment, or *annealing* as it is termed. Diamond was the first gem to be treated by irradiation, at the turn of the twentieth century. Irradiation changes body colour by damaging the diamond crystal lattice, altering its light absorption properties. Irradiation produces a limited range of colour, typically green or blue. Post-irradiation, the diamond may be annealed to partly restore the lattice. Irradiation followed by annealing is capable of a more expanded range of colours: yellow, orange, brown, pink, purple and red-purple. Irradiation treatment is irreversible and heat treatment is probably irreversible. Five techniques have been used to irradiate diamonds:

Radioactive isotope bombardment

Sir William Crookes was the first to subject diamonds to radium. In 1905 he succeeded in permanently colouring diamonds by immersing them in radium-bromide salts for one year. Radium salts spontaneously emit alpha particles, which carry a positive electrical charge. Charged particles are incapable of penetrating the diamond lattice to any depth, thus the irradiation resulted in a surface colouring of blue, blue-green, green and yellow-green. Crookes' diamonds were dangerously radioactive and some of his stones will still possess enough residual radioactivity to register even today on a Geiger counter.

Cyclotron acceleration of proton, deuteron, or alpha particles

In 1942 J. M. Cork used a cyclotron to bombard diamond with accelerated particles, producing a permanent green colour. These diamonds were also radioactive, but only for a matter of hours. The cyclotron is a charged particle accelerator: therefore the resulting colour is only skin-deep. It may be deepened by extended exposure, but it still remains a surface coloration. The shade of green produced is different from that of natural green diamond. If annealed, yellow, orange or brown colours may be produced. Cyclotron treatment produces a characteristic identification feature depending on treatment geometry: diamonds treated in the face-up position through the table show a dark zone of colour at the girdle; diamonds treated through the pavilion show a shape like an opened umbrella when viewed down through the table; and diamonds treated through the side show a zone of colour near the girdle. In addition, brown and yellow annealed diamonds usually show an absorption band at 595nm in the yellow that is not seen with natural greens.

Electron bombardment

Electrons accelerated in a Van de Graff generator are capable of producing blue, blue-green or green colour. All are surface colours because electrons are negatively charged particles. There is no radioactivity and subsequent annealing to a temperature below 400°C produces orange-yellow, pink-purple and brown colours. Identification of blue and blue-green treated diamonds may be done by electrical conductivity: all natural blue diamonds of similar depth of colour are Type IIb and semi-conductors, whereas the treated blues are non-conductors.

Gamma ray radiation

Diamonds treated by gamma rays were first produced in the 1970s. Gamma rays are a type of electromagnetic radiation similar to X-rays, but of shorter wavelength, resulting from the emission of a decaying cobalt-60 nucleus. Gamma rays are capable of penetrating a diamond lattice so the colour occurs throughout the body of the stone; usually a blue to blue-green is produced. The process takes months and the diamonds remain highly radioactive and dangerous to wear. The radioactivity can be detected by a Geiger counter or by production of an autoradiograph: the fogging of a photographic film if the diamond is left in contact with it in a light box. Furthermore, the blue and blue-green diamonds can be identified by their lack of electrical conductivity.

Neutron bombardment

Irradiation with neutrons, uncharged atomic particles, from an atomic reactor is the most commonly used treatment. Neutrons have a much stronger penetration power because they carry no charge and the resulting green colour occurs throughout the diamond. The longer the diamond is exposed to the radiation, the deeper the colour. Any initial radioactivity fades quickly. Subsequent annealing at 500–900°C produces different colours depending on the starting material: yellow, orange, pink, purple and brown can be produced.

Of these five methods only neutron bombardment and electron bombardment are still used today. As a rule, the colours of irradiated diamonds show a narrower range of hue and saturation than naturals and, with some colours in particular, such as dark green and bright blue, the treated colours bear little resemblance to the natural variety. Distribution of body colour can be a useful general observation. Neutron irradiation typically produces a more even colour distribution because the radiation passes through the body of the diamond. Charged-particle irradiation, such as electron bombardment, tends to produce colour zoning that may follow the shape of the diamond. Absorption spectra are the single most useful distinguishing feature, and in most cases give diagnostic results because irradiation and heat treatment usually develop lines or bands not seen with natural colour. Observation of the spectra typically requires laboratory testing with a spectrophotometer at low temperature to increase absorption and define bands.

Irradiation is also used to colour many other gemstones. Gamma rays, electrons or neutron irradiation are used to alter colourless topaz to a brownish colour. If this is followed by heat treatment, a range from light blue, through sky blue to deep London blue can be produced. The colour is stable, undetectable and does not require disclosure. Colourless synthetic quartz that has been ferric-iron-doped can be irradiated to amethyst colour. This synthetic amethyst can be detected by its inclusions: segments of the seed plate may be visible as well as whitish 'breadcrumb' inclusions and also the lack of Brazil twinning seen along the c-axis between crossed polars. Colourless sapphire can be gamma irradiated to produce dark yellow to yellow-brown sapphire. The colour is unstable and usually fades if exposed to sunlight or UV light. Pink may also be irradiated to produce the pink-orange padparadscha colour. With beryl, colourless can be turned yellow; blue can be turned green; and pale pink can be turned the deep-blue maxixe colour. Crème-coloured cultured pearl can be gamma irradiated to a grey-black colour. Colourless zircon can also be irradiated to brown or red. Gamma irradiation of pale pink tourmaline produces a strong pink to red colour that is stable, but there may be some residual radioactivity.

High-pressure/high-temperature annealing

In 1999, Pegasus Overseas Limited (POL) announced plans to commercially market diamonds treated by a high-pressure/high-temperature (HPHT) annealing process developed by General Electric (GE). GE has stated that the process is capable of improving colour, brilliance and brightness of the diamonds and that the treatment is permanent and irreversible. Prior to HPHT technology, permanent colour change in diamond was restricted to additive colour: treatment could alter colour by adding to pre-existing colour, but colour could not be lightened or removed.

HPHT-treated gem-quality diamonds became commercial in 2000, but the process for industrial diamonds has been used by GE since the late 1970s. Type Ia diamonds contain nitrogen in aggregates, either pairs or groups, and Type Ib diamonds, constituting the yellow Cape Series colour range, contain isolated substitutional nitrogen. General Electric's initial HPHT treatment induced aggregation of isolated substitutional nitrogen, thereby converting Type Ib diamonds to Type Ia diamonds and reducing the saturation of colour. The starting material for the current stones, known exclusively by the trade name Bellataire, is Type IIa diamonds, which lack significant nitrogen and are high-clarity, almost pure stones. As such, Type IIa diamonds do not absorb visible light unless they contain defects. The defect of interest is termed 'plastic deformation'. It causes a linear dislocation in the diamond lattice in which a layer of carbon atoms is displaced relative to the adjacent layers of carbon atoms. This occurs post-formation while the diamond is in the mantle or as a result of its violent transport to the earth's surface. HPHT treatment rearranges the defects in the diamond lattice under conditions of temperature and pressure similar to those under which diamond forms in the mantle. This anneals the molecular misalignment and restores the diamond lattice, thereby removing brownish body colour.

The rarity of Type II starting material dictates that only a few thousand stones suitable for treatment are available every year; however, they are premium goods: approximately 60% are IF or VVS_1 quality and approximately 40% are in the VVS_2 to VS_2 range. Type IIa diamonds are characteristically irregular crystal shapes, and thus almost all of these stones are cut as fancy shapes. HPHT treatment makes dramatic changes in colour: of diamonds in the N–O range through to Light Fancy brown, most are improved to D or E colour.

Bellataire diamonds are sold through select retailers and every stone is laser inscribed on the girdle and comes with a GIA grading report stating its quality and the processing. The laser inscription is clearly visible with 10X magnification but because it can be polished off it is not a foolproof guarantee of disclosure. Some general observations that can induce suspicion include: graining which is typically more pronounced and which may exhibit a whitish appearance, as well as a slightly hazy appearance to the stone. At present there is no single property that distinguishes HPHT treated diamonds, and specialized laboratory equipment is required because the best general identification for all HPHT decolorized diamonds are the features seen in the infrared and photoluminescence spectra; specifically, the most important spectra are those of laser-excited photoluminescence (PL) spectroscopy as seen under low-temperature analysis.

The first step in identifying a potential HPHT treated stone is to determine whether the diamond is Type II. The SSEF Diamond Spotter developed by the Swiss Gemmological Institute distinguishes diamond based on the transparency of Type II diamond to SWUV light. Alternatively, DeBeers DTC's Gem Defence Program developed the DiamondSure™ to detect the presence of the Type Ia Cape spectrum 415nm band found in 95% of all natural diamonds. Its absence indicates either synthetic origin, the original purpose of the instrument, or D and E coloured stones, some fancy colours as well as Type Ib and Type II. Use of these instruments can effectively reduce the number of diamonds that have to be submitted for testing. Once a diamond has been determined to be potential Type II it should be submitted to an accredited laboratory where all diamonds are routinely tested for evidence of HPHT and its presence is reported on all certificates.

Similar HPHT conditions have been applied to successfully induce fancy colours. The resulting colour depends on the Type of diamond being treated and the parameters of the HPHT process. The colour most often produced is yellowish-green or greenish-yellow but pinks, blues, oranges and yellows in various shades are also produced. Again, identification requires laboratory testing, but as general observations the majority show a chalky fluorescence and the body colour is typically highly saturated, most often with dark tones.

Fracture filling

Fracture filling is the process of infusing a specialized glass compound into surface-reaching fractures to improve the overall appearance of diamond. The process, invented by Zvi Yehuda in 1982, likened by some to filling cracks in a windshield, is capable of visibly minimizing certain open inclusions yielding a diamond with more life and fire. Fracture filling is commercially available worldwide on an extensive range of goods, including coloured diamonds. Typically I_2 and I_3 stones are filled but stones in the SI grades are also treated, as well as a few VS.

When light travels into a diamond with a large fracture, the fracture obstructs the path of light because the air that resides within the walls of a fracture has a much lower refractive index than diamond, which causes the fracture to become visible in high relief. Filling the fracture with a stable compound of high RI, one that approaches the 2.417 of diamond, reduces the differential in RIs and permits light to pass through the fracture with less disruption, thereby allowing the diamond to exhibit more brilliance. Air is vacuumed out of the fracture and the filling compound is introduced in one continuous process over a brief interval without interruption under a high heat of at least 400°C at a pressure of approximately 50 atmospheres. Stones ranging from 0.01ct to over 50.00ct have been successfully filled, but the typical commercial range is 0.30ct to 5.00ct. The durability of any fracture-filled diamond is dependent upon factors exclusive to that stone, including the chemical composition of the filler and the physical dimensions and nature of the fracture. Proceed with repairs and cleaning according to Yehuda's rule that 'anything that can be done with an emerald may be done with a Yehuda diamond'.

All fracture-filled diamonds are identifiable by microscope with requisite knowledge. Often the laser inscription of the treatment firm is visible, but sometimes it can be polished off or obscured by the setting. Systematically examine the stone while gently rocking it back and forth to increase the viewing angles. The most obvious identification feature is the characteristic colour flash, an internal reflection caused by the difference in dispersion between diamond and the filling compound. Flash effects are more easily detected if seen parallel to the fracture and are almost always seen through the pavilion. Gas bubbles trapped within the filler may also be seen and whitish cloudy areas, typically near the surface, sometimes reduce filler transparency. Cracks in filling material can appear similar to mud cracks in a dry lakebed. In addition, X-radiography is capable of revealing the presence of filler in most treated stones because filled areas appear distinctly white on the X-ray film in sharp contrast to the extreme transparency of diamond. X-ray fluorescence also shows a difference between the filling compound and the carbon of the diamond. Disclosure of fracture filling is required by jewellery industry standards. The International Diamond Manufacturers' Association (IDMA) and the World Federation of Diamond Bourses (WFDB) passed resolutions that major labs should refrain from grading filled diamonds because true colour and clarity cannot be determined with the filler present and it is not necessarily a stable permanent treatment.

Glass infills had been used on other stones decades before the diamond treatment was developed. It is common for faceted ruby and sapphire to have surface pits that are not polished off due to weight retention issues, and occasional fissures that reach the surface. These pits and fissures can be glass infilled to improve the appearance and surface lustre of the stone. It is often possible to see the difference in reflectivity between the corundum and the glass infill material.

Laser drilling

Lasers produce intense beams of in-phase, precisely parallel light waves that can be focused to a fine point capable of the directed structural alteration of dark inclusions in diamond that are situated too far below the surface to be economically removed during polishing. The laser drilling of diamond is a two-stage process that starts at the dark inclusion and progresses out towards the surface of the stone.

In a high-temperature/high-pressure environment, the stable crystal structure of carbon is diamond. In the lower-temperature/lower-pressure environment in which we live, the stable structure for carbon is graphite because this structure uses less energy, with one of the four carbon bonds being much weaker.

The laser beam focused on the dark inclusion generates enough heat to allow the structure of the carbon bonds to alter to the lower-energy structure of graphite. Even though one in four carbon bonds converts to the weak bond associated with that across the cleavage direction of graphite, there is no room for the material to expand to the lower SG of graphite. The diamond-converted-to-graphite then has the bonds of graphite in a structure distorted by the pressure of its enclosure within the surrounding diamond. As the focus of the laser is pulled out towards the surface of the diamond, a graphite pillar is left behind where the diamond has been converted to graphite by heat in the oxygen-free interior of the diamond. When the graphitization reaches the exterior, oxygen at the surface can combine with the carbon of the graphite pillar to form CO_2 that will vaporize away. Graphite that remains in the drill tunnel is removed by a process called sparking; because graphite is an electrical conductor, a high-voltage high-frequency coil can induce a current in the graphite that heats it to a combustion temperature and burns away the graphite column, leaving a characteristic feature known as a drill tunnel or a drill hole. Early pulse-style lasers left ragged tunnels that were relatively easy to detect; drilling is now performed by continuous wave (CW) lasers, which produce smooth, clean drill tunnels of approximately 15–20 microns.

Suitable inclusions are dark piques because lasers must be focused on dark objects in order for sufficient heat to be produced; graphite and metal oxide inclusions can be removed but not inclusions consisting of many small piques or of a crystalline nature. The process replaces high-relief black inclusions with colourless inclusions that can dramatically increase the attractiveness of a diamond, although it very seldom raises the clarity grade. Typically, an SI_2 with a black inclusion is improved to a better-looking SI_2 with a colourless inclusion. Laser drilling, also invented by Zvi Yehuda, became a commercial diamond treatment process in 1972 and is accepted by the trade, if disclosed.

Systematic examination with a microscope is required to see CW drill tunnels. Drill tunnels are typically seen through the pavilion and may be confused with two natural features: etch channels and rutile needles. Natural etch channels have square, triangular or hexagonal cross-sections, are rarely straight, and because they follow atomic-size crystal defects, they appear to meander randomly within the diamond. Laser drill tunnels have round or oval cross-sections and are almost always straight. And, unless it has been missed, laser drill tunnels always lead from an inclusion recognizable as a whitish void or a colourless fracture that may or may not be fracture filled to reduce relief. Check the surrounding area of all such inclusions carefully. On rare occasions, fine needle-like rutile inclusions are found in natural diamond; however, the needle-like inclusions most likely to be confused with laser drill tunnels are those found in moissanite, so be certain the stone is in fact diamond. Over-exposure during the sparking process can leave small burnt areas on a single facet. Such areas occur only on the surface at the drill tunnel. Burns may also result from the cutting process, but these typically extend to a larger area covering more than one facet.

In early 2000 a new lasering technique referred to as KM entered the trade. Instead of producing a characteristically identifiable drill tunnel, KM treatment employs one or more pulse laser beams to vaporize an inclusion and then create sufficient stress combined with internal pressure generated by the vaporized inclusion to broaden already existing sub-surface fractures, or create new internal fractures, and extend them to the surface, leaving a more natural-looking feature. The inclusion is bleached by pressure-boiling the diamond in acid. Pressure is necessary because the conduit is so narrow that, without it, the acid would not be able to penetrate the stone. Ideal candidates for KM treatment are diamonds with shallow dark inclusions accompanied by internal fractures. The deeper the inclusion the less suitable it is because its treatment would require much greater internal pressure and leave a long fracture that might be more noticeable than the original black pique. Most KM-treated diamonds are 1–2ct, but stones as large as 5ct have been treated. It is not known how prevalent the treatment is in the trade.

KM treatment can be difficult to detect and systematic examination should be done under microscope magnification. There is no single distinguishable feature, but the feature seen most often is a feature unique to the boiling process: a flash effect somewhat similar to that exhibited by fracture-filled stones. The difference between the two is the colour range: boiled diamonds tend to show interference colours changing from light blue to light brown when the stone is tilted back and forth.

When laser drilling became a commercial treatment in the early 1970s, it was subject to a requirement of disclosure. But for a time thereafter, disclosure was not required by: the United States Federal Trade Commission (FTC), the World Federation of Diamond Bourses (WFDB), and the International Diamond Manufacturers' Association (IDMA). The International Standards Organization (ISO) was in favour of disclosing laser drilling, and GIA as well as the Jewelers Vigilance Committee (JVC) lobbied the FTC and international trade organizations to mandate disclosure of laser drilling. The non-disclosure policy has since been reversed; however, it is important to understand that diamonds transacted during the non-disclosure period may not be properly reported as such on older certificates. At present, accredited laboratories grade laser-drilled diamonds because the treatment is permanent and without durability issues. The laser-drill tunnel is considered as an inclusion and graded accordingly: plotted as an inclusion and listed in the key to symbols section of the plotting diagram.

LESSON 16

Gemstone Cutting and Styles of Cut

When gemstones are found in nature they may have distinctive shapes with plane and lustrous faces; sometimes these crystals are mounted in jewellery as they are, but more often than not the faces and edges of a crystal have suffered attrition during their transport to the surface of the earth and to the deposit in which they are finally found. Over these countless years they may be completely abraded, leaving only a simple rolled or water-worn pebble. The beauty and symmetry of gemstones we see in jewellery is a product of the art of the gemstone cutter, or lapidary, as they are called in the trade. Lapidaries cut all species of gems whether common, rare or ornamental, except diamond. Diamond cutting is a specialized craft and those who engage in it are simply called 'diamond cutters'. Diamond cutting differs from the fashioning of all other gemstones in two ways: diamonds can only be cut with their own powder, and the forming of the facets and their polishing is carried out on the same rotating wheel in one operation. In the cutting and polishing of other gemstones, separate processes are carried out.

Diamond cutting

The process of transforming a rough diamond into a polished shape is carried out by a number of operations: designing and marking; dividing – cleaving or sawing; bruting; and faceting – cross-work and brillianteering. All traditional manufacturing is a diamond-on-diamond operation.

The designer studies the rough and determines how to produce the largest, cleanest diamond(s) of finest make – in other words, how to maximize value from the rough. To this end, the designer decides the following: the number and shape of diamonds to be cut; how to divide the rough; and how inclusions should be removed or incorporated into the stone. Cutting only one stone from most rough results in too much waste. Regular rough, particularly the octahedron, is almost always fashioned into two round brilliants, typically one larger and one smaller as this has the greatest market value in most instances. Fancy shapes are typically cut from irregular rough. The weight of fashioned diamonds taken as a percentage of original rough weight is termed 'yield'. When cutting a well-shaped octahedron into two parts that will become two round brilliants the mean value for yield is 52%. The stone is marked with a very fine black pen to indicate the tip of the culet and the plane on which the stone is to be divided by sawing or cleaving. Diamond cannot be divided or polished in just any random direction as it cannot be cut in its hardest direction or sawn or polished on a cleavage plane, so the stone must be marked in accordance with crystallographic requirements, specifically the grain. Also, wherever possible, the stone is oriented so that inclusions can be removed during dividing, bruting or faceting.

Cutters refer to the harder and softer directions relative to crystallographic orientation as the grain. The softer direction is 'along the grain' while the harder direction is said to be 'against the grain'. Grain is established by crystal growth and is present in every diamond regardless of its external shape. An octahedron may be sawn on the four-point and two-point grain, and cleaved on the three-point grain. The chosen grain direction also dictates the eventual polishing directions. The four-point direction, the most common way of sawing an octahedron, is parallel to a cube plane. This orients the table parallel to the natural girdle of the octahedron. There are four grain directions. This is often misunderstood to be two grain directions, but in fact the reverse of one grain direction is another grain direction; therefore when the grain runs parallel to the sides of the octahedron forming a cross on the cube face, the forward and reverse along two intersecting lines gives four grain directions. The two-point direction is parallel to a dodecahedral plane. In this case the table would bevel an edge of the octahedron. A two-point stone has two grain directions that run parallel to the length of the face. Again, the reverse of one grain direction is another grain direction:

towards one end is one grain direction and towards the other end gives the second grain direction. The three-point direction, the least popular of the orientations, is parallel to an octahedral face. This results in four directions of cleavage corresponding to the four pairs of parallel faces on the octahedron. A three-point stone has three grain directions forming a triangular shape.

The decision as to whether to saw or cleave a diamond to divide it depends on its shape, inclusions and the grain. Cleaving is the act of dividing a diamond into two along a cleavage direction by means of a sharp blow – likened to splitting wood along its grain. Cleaving is favoured to quickly shape distorted rough or to divide large diamonds. Cleaving is an ancient art employed far less frequently since the development of the diamond saw because cleaving is restricted to the four directions parallel to the octahedral faces, making it impossible to cleave an octahedron into two for the traditional manufacturing of round brilliants.

Sawing is the process of dividing diamond in a non-cleavage direction by cutting – that is, scratching or abrading – through it with a blade charged with diamond dust. All well-formed crystals are typically sawn because sawing directions will provide better yield than cleaving directions. An octahedron can be sawn so that the saw cut automatically produces a table: through the centre typically parallel to, but just above, the natural girdle of the stone. This produces two unequal pieces yielding a larger and a smaller round diamond. Sawing machines consist of a cast iron bed that holds the sawing disc assembly in the front and a movable weighted arm at the back. The diamond is cemented into a *dop*, a generic term for any device holding a diamond during manufacturing, and clamped into the saw. Saw blades are porous phosphor-bronze discs, the edges of which are charged with diamond dust. The diamond is lowered on to the blade rotating at 7000–15 000 rpm and sawing continues until the blade has cut the diamond completely in two.

Bruting is the process of producing the girdle outline in preparation for faceting by using a stationary diamond to grind away material from a rotating diamond. It is also often called girdling, rounding up, or simply cutting. It is only used on diamonds with rounded outlines such as round brilliants, oval brilliants, etc. Emerald cuts and other stones with straight girdles are never bruted. Bruting machines are similar to woodworking lathes. The diamond is cemented on a dop that fits into the rotating chuck of the machine. The sharp, which is another diamond in line for bruting, is mounted in another dop attached to the end of a long stick. The bruter holds this stick under the arm for leverage, presenting the sharp's hardest direction upward, and the handle is placed in a U-shaped support at the front of the machine. The sharp is touched to the rotating stone from the bottom, and grinding proceeds by moving the sharp from side to side across the girdle of the rotating stone; this side-to-side motion keeps the girdle straight and flat. The more skilled the bruter, the more perfect the girdle outline and the greater the weight retention. The presence of a small portion of the original crystal, called a *natural*, is proof that maximum diameter, and therefore maximum weight, has been retained.

Faceting is essentially two identical operations performed one after the other: grinding and polishing. Grinding is a coarser operation that involves forming a facet, leaving a surface that has fine lines on it. Polishing is the act of smoothing the facet to remove these lines. Manufacturers call both these operations faceting; the distinction is only that of surface finish. Both operations are done on the same machinery: a simple horizontally mounted wheel, the *scaife* (pronounced to rhyme with 'life'), which revolves like a record turntable. A scaife is a cast iron disc about 30cm in diameter and 2.5cm in thickness. It is dressed with a paste of diamond dust and olive oil. The diamond is held in a dop and secured to a supporting tang used to direct the diamond into position on the scaife. In essence, the tang is a tripod with two broad legs at the back that rest on the bench, and the third leg is the diamond on the scaife. Adjusting the dop sets the precise angle.

Diamond is faceted in two stages: cross-work and brillianteering. The cross-worker puts on the first 17 facets. The table is formed first, followed by the four corners, or *quoins*, leaving the table shaped like a perfect square. The crown facets are now subdivided with four bezel facets, which are polished on the ribs of the four corner facets transforming the table to an octagonal shape surrounded by eight crown main facets. The same is done on the bottom. The stone is now said to be single cut. Today, a culet is rarely cut. At this stage the bruter gives the girdle a final polish, termed the *rondisting*, before the brillianteerer completes the remaining facets.

The brillianteerer adds 40 facets: 24 on the top and 16 on the bottom, all of which are triangular in shape. The eight star facets are formed first and are placed at the junction of the edge between two of the eight crown main facets and the table. Each star facet extends halfway towards the girdle. Next are the 16 upper girdle facets, or halves, two for each crown main facet, followed by the 16 lower girdle facets, or bottom halves, which are polished directly below the upper girdle facets and extend about 75–80% of the distance from culet to girdle. When the brillianteerer is finished, the stone is full cut. Polishing or faceting the girdle is optional and done as a last step. Smaller brilliant cut diamonds are often left with a matt girdle since the condition of the girdle does not influence brilliance to any significant degree. Reflections of the girdle cannot be seen through the crown of a well-proportioned brilliant.

Diamond cutting is a traditional craft; however, the last twenty years have seen technology applied selectively in certain areas: computer imaging techniques for assessing optimal cutting yield from rough, laser sawing, laser kerfing, and automatic bruting and polishing. Such developments apply more to large operations; most diamond manufacturing is still performed in the traditional way.

Gemstone cutting

Unlike diamond, which is faceted and polished in one operation, with stones of other species there must be two distinct operations: that of cutting the stone to the required shape, which leaves the facets with a matt surface, and the polishing of these facets to a mirror-like surface.

A piece of rough gem material is first *slabbed* into suitable-size pieces by sawing it on a vertically mounted diamond or Carborundum®-charged disc. This is similar to the sawing of diamond except that in this case the material is hand-held during the sawing process. These pieces are marked with the intended cut style's girdle outline, and unwanted material is removed with a trim saw and/or by grinding. Whether or not the grinding extends above and below the girdle to begin to shape the profile, this interim shape is the *preform*.

The rough is then taken to the cutter who, using a cast-iron, copper or gunmetal horizontal lap, grinds on the facets. Diamond dust, Carborundum® or emery is used as the abrasive. The angles of the facets are controlled by the angle of the *gem stick*, into which the stone is cemented, and the *gem peg*, which is a vertical post drilled with a number of holes into which the end of the gem stick can be inserted to achieve the angle. Afterwards the stone is thoroughly cleaned so that no abrasive is left on the surface, and then transferred to another bench for the *polisher* to polish the already-ground facets. This is carried out on copper, pewter, wooden, leather and even plastic laps, with polishing agents such as jeweller's rouge; rottenstone; cerium, tin or chromium oxides; or graded corundum and diamond powder. The use of mechanical stone holders has to some extent simplified the work of gem cutting, but the best work is still carried out by hand.

The body colour of a coloured stone is almost never uniform and it is the lapidary's job to orient the stone to produce the best colour through the table. With stones that exhibit pleochroism, such as ruby, sapphire, tourmaline, etc., care has to be taken to ensure that the table is placed in the correct orientation to display the best colour, sometimes meaning the various colours, in the face-up position. Stones with chatoyancy and asterism, cat's eyes and star stones, must be oriented with the needles or cavities parallel to the base of the cabochon. With stones that have pronounced colour zoning, as does sapphire, the location and orientation of the colour zones within the finished stone can have a dramatic influence on the appearance and beauty of the stone. Lapidaries achieve best colour by setting the orientation of the crystallographic axes, the location of colour zones, and also by altering facet angles and proportions. This is the reason that cutting styles used for coloured stones have a greater number of variations and fewer regular facet angles and configurations than those used for diamonds. Cutting styles for coloured stones have to have more flexibility in terms of lengthening the path of light through the stone, or alternatively shortening the path of light through the stone, to deepen or lighten colour. Other variables are also manipulated: thick girdles and crowns produce deeper colours, and larger tables produce lighter colours. It is common to have sides that are cut to different angles and asymmetrical faceting is also frequently applied. Lasers are now employed to cut coloured stones, and with this method any shape, including those that are concave on the bottom or sides, can be cut; this was formerly an impossibility with traditional lapidary equipment. Cylindrical drums (replacing flat laps) can be used to sand and polish concave facets.

The engraving of cameos and intaglios is carried out by the use of steel blades, files and small rotary burrs fitted into a drill chuck on a drill head or flexible drive. Holes may be drilled by needles charged with diamond powder or other abrasives or by similarly charged tube drills. Tumbled gemstones are produced by grinding rough gem material to an irregular shape in a rotating drum using a grinding powder; and then similarly polishing the stones in another drum using a polishing powder.

Styles of cut

Cut style refers to the geometrical pattern of facets of a finished gem, the style into which a gemstone is fashioned. To exhibit the maximum beauty of a gemstone, certain definite shapes and angular arrangements are beneficial: this is particularly so with diamond and with coloured stones having special considerations such as pleochroism, colour zoning or a colour in need of deepening or lightening. Such considerations have produced a series of styles of cut that are most suitable to each particular type of stone. To this day, many cut styles remain primarily dependent on taste and demand, and in recent decades there has been a renewed interest in antique cuts: rose cuts and briolettes in particular.

Diamond cut styles

The first known cutting style was the *point cut* fashioned by polishing facets to mimic the faces of the octahedron. It emerged near the end of the fourteenth century when a widely held belief was that diamonds must be unaltered to retain their mythical powers. Thus, fashioning improved symmetry and transparency while keeping crystals as close as possible to their original shape.

The *table cut* was produced by grinding away the top point of an octahedron, perhaps a point that was broken or abraded, leaving a truncated square facet forming the table, and the four sloping sides of the octahedron produced the crown. Table cuts dominated throughout the sixteenth century and into the early seventeenth century. Those that were cut with an oversize culet, termed *tablet cuts*, were usually foiled to compensate for it. Extremely shallow tablet cuts, consisting of a table facet and a culet separated narrowly by a girdle, were termed *portrait stones* because they were used to protect miniature portraits. In the *stepped table cut* the four sloping edges on the crown were fashioned to produce four steps. Eventually both the crown and pavilion came to be stepped. The stepped table cut is the predecessor of modern step cut styles.

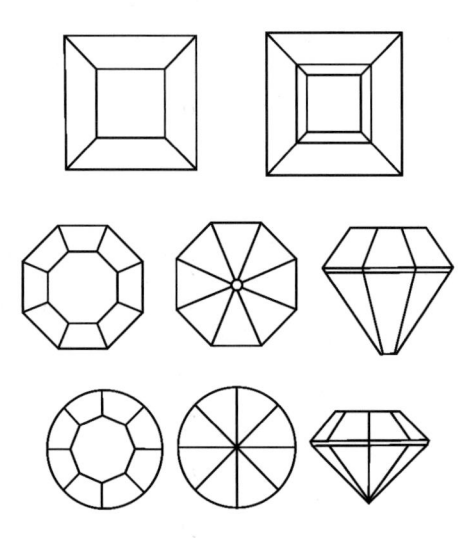

Figure 16.1 Diamond cuts. Top left: table cut. Top right: stepped table cut. Middle: old single cut. Bottom: modern single cut.

The first *single cuts* evolved from modifications of table cuts with broken corners or undesirable inclusions. The natural sloping sides of the octahedron create the first four crown facets, and grinding down the corners forms four additional matching crown facets. This was repeated on the pavilion for a total of 18 facets: 8 on the crown, 8 on the pavilion, plus a table and a culet. The old single cut is similar to the modern-day single cut except that the modern stone is not as deep and its outline is circular. Only the modern single cut was produced, and still is produced, in quantity.

The *double cut* emerged circa 1620. There were 34 facets in total: 16 on the crown, 16 on the pavilion, plus a table and a culet. A double cut with a cushion-shaped girdle outline is often referred to in literature as the Mazarin cut, being falsely attributed to the famous diamond collector Cardinal Mazarin.

By the mid-1700s, the *triple cut* had evolved and increased the number of facets to 58 in total: 33 on the crown, including the table, and 25 on the pavilion, including a culet. It differs from modern round brilliants in that the outline was squarish to maximize weight retention of octahedral rough and it had different depth, crown and pavilion angles, table size and culet size. Despite these differences, this cut is credited with establishing more symmetrical size and facet angles capable of producing a marked increase in brilliance and fire over the double cut. The triple cut is often referred to as a Peruzzi cut. This is another false attribution from a mistranslation of a book published in 1826 and there is no confirmed record of Vincenzio Peruzzi or a Peruzzi cut. The triple cut is also referred to as the *Old Mine cut*.

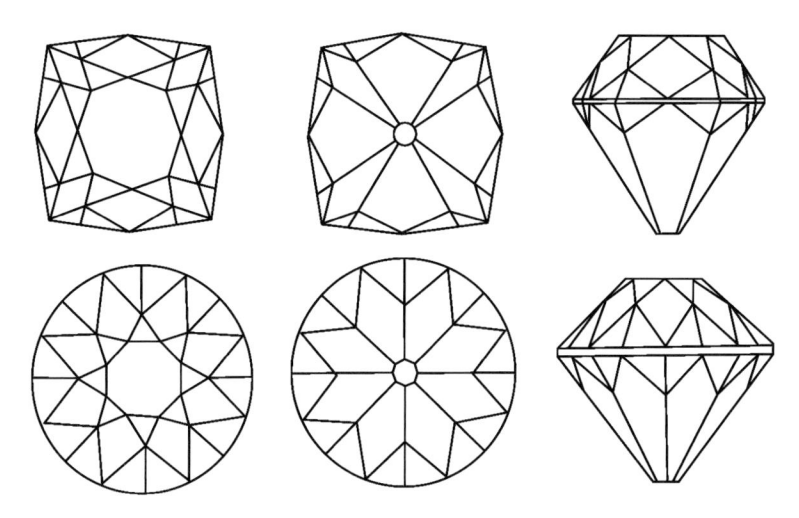

Figure 16.2 Diamond cuts. Top: Old Mine cut. Bottom: Old European cut.

The *Old European cut* fundamentally retained the proportions of the Old Mine cut, but was modified with a circular girdle, thus creating the first true 58-facet round brilliant. In reality, because weight retention was the primary consideration, most girdle outlines were cushion-shaped. This changed by 1820 when mechanical bruting was developed and it became possible to round up a girdle to produce a circular outline. Nonetheless, the Old European cut retained its greater depth and large crown and pavilion angles because of weight considerations.

The first modern *round brilliant* is attributed to Henry Morse, who by 1870 was producing cuts with angles and proportions near to the preferences of today. By 1900, with the advent of the mechanical saw, his shape and proportions became popular. Before a stone could be sawn, the point of an octahedron had to be ground away to form the table. This meant that a thicker stone achieved maximum weight retention. The invention of the saw, allowing diamond to be sawn parallel to the three possible girdle planes of the octahedral rough meant that adopting a thinner crown and larger table retained the greatest weight. Around this time efforts were being made to mathematically compute the most desirable angles for optimum brilliance, dispersion and scintillation. With this cut, it is most important that the observer see an attractive pattern of reflected light evenly distributed throughout the stone when looking into the crown. Total internal reflection occurs when light strikes the back facets at greater than the critical angle, but small variations with the limits imposed by the critical angle can dramatically alter the attractiveness of the reflection pattern.

Marcel Tolkowsky published the results in his 1919 book entitled *Diamond Design*. Tolkowsky's ideal had a 53% table, 59.3% depth, 16.2% crown height, and a 43.1% pavilion depth. Since Tolkowsky's time, the proportions of the modern round brilliant have changed slightly, the principal differences being a larger table, a decrease in crown depth, and extended lower girdle facets. The modern round brilliant cut has 57 facets in total: on the top, or *crown*, of the standard round brilliant there is a large central eight-sided *table* facet, which is surrounded by eight triangular *star* facets. There are eight *crown main* facets, alternatively termed bezel or kite facets, and 16 triangular facets referred to as *upper girdle* facets bordering the *girdle* of the stone. On the bottom below the

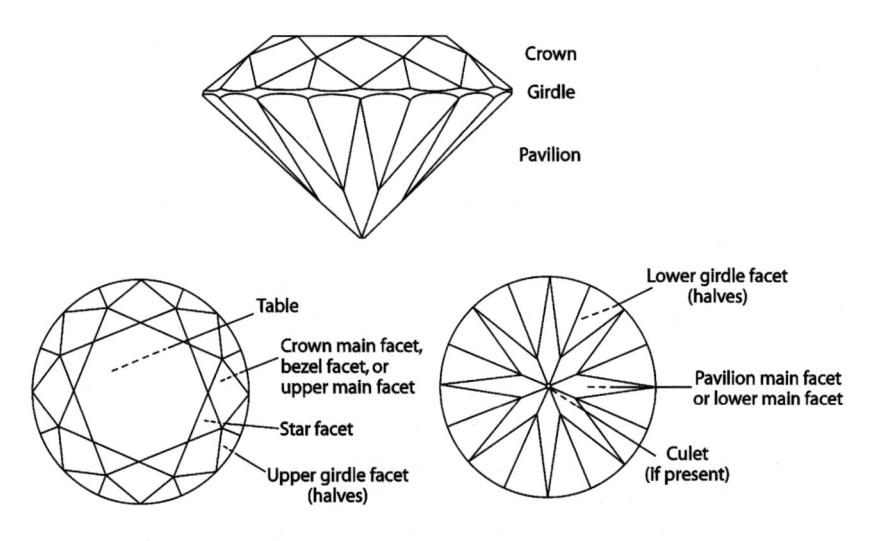

Figure 16.3 The round brilliant cut.

GEMSTONE CUTTING AND STYLES OF CUT

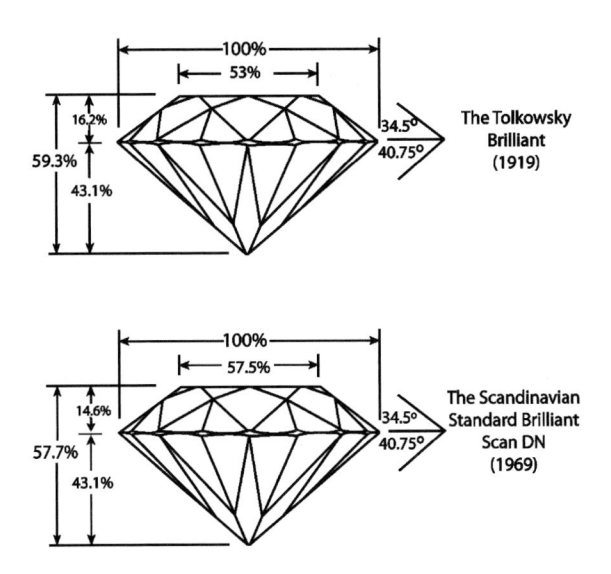

Figure 16.4 Ideal round brilliants.

girdle, on the *pavilion* of the stone, there is a similar set of 8 *pavilion main* facets, and also 16 triangular facets referred to as *lower girdle* facets bordering the girdle of the stone. Sometimes the round brilliant is said to have 58 facets but since a culet is rarely cut today, it is convention to refer to the modern round brilliant as having 57 facets.

Fancy shapes

Fancy shape refers to any modern cut diamond other than single cut or round brilliant. These include both fancy brilliants and modified brilliants.

Each facet of a *fancy brilliant* corresponds directly to a similarly positioned facet on the modern round brilliant; however, these matching facets cannot be all the same size, as a result of distortion caused by the non-round girdle outline. None of these varieties equals the brilliance, fire and scintillation of the round and their light efficiency reduces proportionately with departure from the round outline. The four traditional varieties are the oval, pear, heart and marquise. Alternative varieties include: cushion-shaped brilliants, commonly used for fancy-colour diamonds, and shapes based on the triangle.

Modified brilliant is a term used to describe cuts with outline shapes and/or facet configurations similar to, but not corresponding directly to, those of the round brilliant. There are countless variations including triangles, octagons and cushions. The two most commercially successful are the *princess cut* and the *radiant cut*. The princess cut has a square or, less often, rectangular outline and the corners may be truncated. The design evolved to increase the brilliance of square and rectangular cuts by increasing the number of facets on the crown and pavilion. There is no standardized arrangement, although most have 76 facets. The princess cut's most desirable feature is that it achieves greater weight yield from octahedral rough than a round brilliant. The radiant produces the optical effects of a brilliant from a cut-cornered square or rectangular stone. There are various facet combinations with no set standard and the average number of facets is 70. It is also advantageous in terms of yield. The disadvantage of many of these square, rectangular or triangular shapes is that the sharp corners tend to suffer breakage.

Rose cut

Rose cut facet configurations originated in India and were in wide use by the middle of the fifteenth century, but the name 'rose cut' was not actually used until late in the seventeenth century when the facet arrangement was equated to the petals of an opening rose bud. The traditional rose cut is flat on the bottom and dome shaped on top. It consists of a series of triangular facets rising from a flat base to a point at the apex. It was ideal for small cleavage fragments and macles of diamond, and was also a popular style for pyrope garnet used in Victorian jewellery. Rose cuts may be fashioned with any of 3, 6, 12, 18 or 24 facets. In the full rose cut, the top is two-stepped with triangular facets of 6-fold symmetry that terminate in a point – typically 6 on the first step, and 18 towards the base. The *double rose cut*, domed and faceted on both sides, was popular in the nineteenth and early twentieth centuries for earrings and watch chain pendants. Rose cuts were very important in the jewellery of their time and a number of famous diamonds, including the Great Mogul and the Koh-I-noor, were cut in this style. After 1800, the style was probably not cut except by special order; however, as of the early 2000s, Indian cutters were again fashioning rose cuts.

Beads and briolettes

Beads and briolettes are styles of cut in which the stone surface is covered with an indeterminate number of small facets, most often in steps, and the girdle is absent. The facets are usually triangular shaped, but square and oblong facets are also used. Beads can be spherical, elliptical, or shaped like a barrel or cylinder. They were an ancient form of fashioning

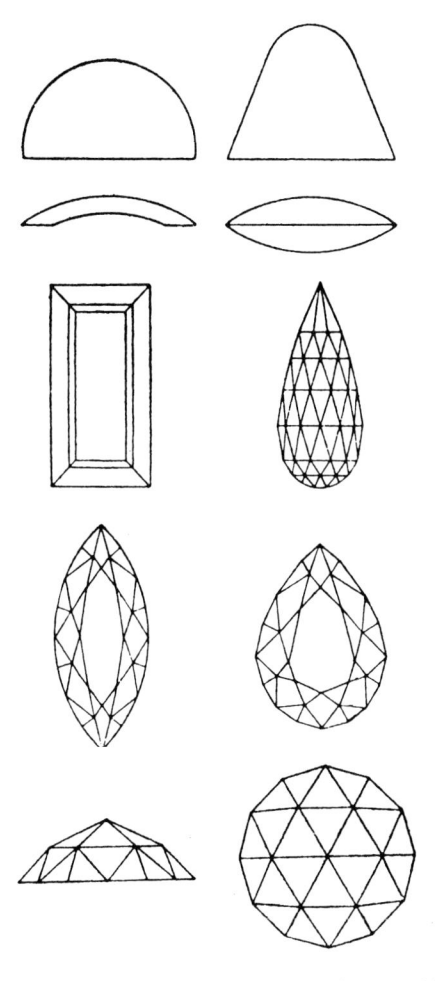

Figure 16.5 Cut styles. Top row: medium cabochon and high-top cabochon. Second row: hollow cabochon and double cabochon. Third row: straight baguette and briolette. Fourth row: marquise brilliant and pear shape brilliant. Bottom row: rose cut – profile and top view.

for many gems. Briolettes are essentially pear- or drop-shaped beads with one pointed end. Historically these stones were described only as *pendants* or *pendeloques*. As a cutting style for diamond, the briolette was popular in olden times, but suffered a loss of favour when the pear shape brilliant appeared; however, since the late 1990s, the briolette has experienced renewed interest and is reclaiming popularity among the elite as a cut of interest for pendants and drop earrings. It remains popular for other transparent faceted stones.

Cabochon cut

This is the simplest cut for a gemstone. It consists of a curved, or domed, upper surface which may be either low, medium or high-topped, on a base which may be flat – creating a simple cabochon; convex but typically less curved than the top – forming a double cabochon; or hollowed out – creating a hollowed cabochon. The outline of a cabochon is commonly circular or oval, but other outlines are also seen such as those that are pear shaped. The style is most suitable for translucent and opaque stones, transparent stones that are too highly included to be faceted, opals, and stones that have particular optical effects such as cat's eyes and star stones. This cut is also used for almandine garnet, in which case the back of the cabochon is hollowed out to lighten the colour; such a stone is referred to as a *carbuncle*.

Zircon cut

This cut has a similar facet configuration to the brilliant cut except that an additional set of facets is added to the pavilion; these reach from the culet halfway up the back facets. As its name indicates, this cut is often used to give life to pale blue heat-treated zircons and add confusion to the internal reflections and so mask any obvious doubling visible through the crown.

Step cuts

The step cut, alternatively known as the emerald cut or trap cut, refers to a style characterized by long parallel facets. The emerald cut is a rectangular shape with cropped corners and tiers of long facets. Typically, emerald cuts exhibit 58 facets, including 8 that comprise the girdle. This is the cut most often used for emerald, and occasionally for diamond – in which case it produces reasonable brilliance but lacks fire. Many other step cut shapes have been fashioned including: square, triangular, kite, hexagon and others. One such variation is the baguette, designed exclusively for complementary stones, termed side stones. It is an elongated cut with a typical length-to-width ratio from 2:1 to 3:1. Shapes and proportions vary considerably and baguettes are produced both straight sided and tapered. They are ordered by millimetre measurements but paid for by carat weight. A modification of the step cut consists of the long rectangular facets being divided into four triangular facets. This is called the *scissors cut* or the *cross cut*, and it is often used on larger sizes of synthetic corundum or synthetic spinel as well as on glass simulants.

Mixed cuts

This cut, which is used occasionally for fancy colour diamonds but more typically for coloured stones, combines both brilliant-cut and step-cut faceting. It is possible for either a crown cut in the brilliant style to be atop a step-cut pavilion, or for a step-cut crown to be superimposed on a brilliant-cut pavilion. When used for diamond, a step-cut crown is typically superimposed on to a brilliant-cut pavilion. Facet configurations do not usually conform one-to-one and there is often a great deal of variation. Many shapes are produced: rectangles of both octagonal and cushion outline, triangles, squares and various others.

Additional terms

A *cameo* is a stone with a raised carved image; if the carving is engraved into the stone, as for the purposes of a seal, the stone is termed an *intaglio*. A *curvette* is a cameo engraved so that the design has a hollowed background with the edge of the stone raised at least as much as the central design.

Gemstones are weighed in *carats* and typically ordered by carat weight or by size in millimetres. *Melee* is a term used for small diamonds and *mélange* is the corresponding term used for a mixture of larger sizes. The term *calibre cut*, which means 'measured', is used for stones 'cut to measure' for incorporation into a special jewellery design. It is used for diamonds and also to refer to small step-cut stones.

LESSON 17

The Pearl

The earlier lessons of this series gave accounts of gems and imitations that are products of the mineral constituents of the earth, or in the case of the imitations, products of scientists and technologists. Still, there is one group of gems that have as their genesis living creatures; that is, their origin is organic and not mineral. These gems are the subjects of the next two lessons. Chief among organics is that prized possession known as the pearl. It was one of the first gems worn.

What is pearl and how is it formed? Beyond the common knowledge that pearls are found in 'oysters', little other information is generally known. Practically all shelled molluscs have the ability to produce substances that together create pearls and mother-of-pearl, but it is only in certain types that gem-quality nacreous pearls, pearls as we know them, are formed. Natural pearls form without human intervention. They are exceedingly rare in the marketplace today, although for many centuries this was the only type of pearl known. Cultured pearls form with human intervention and are cultivated on a pearl farm. Almost all pearls on the market today are cultured.

In this context 'oyster' is in fact an inaccuracy. The animal that produces pearls is not strictly a true oyster, at least not of the same zoological family, *Ostreidae*, as are the succulent morsels so well known to the gourmet. Pearl oysters belong to the distinct order *Pterioida*. These are molluscs. Pearl-producing molluscs are bivalves, meaning their shells have two halves connected by a hinge, like a clam. There are approximately 30 000 species of bivalve molluscs, but only a few are known for pearl production. These have the ability to secrete crystalline calcium carbonate and an organic substance known as *conchiolin* in order to produce the hard covering, or shell, which serves as the protection for the soft body of the animal. And it is this secretion that will, in certain conditions where damage or injury causes irritation to the animal, produce the various types of pearls. Hence, all pearls are an abnormal condition. There are two main types of pearls. *Cyst pearls* are pearls found inside the body of the animal and these constitute the finest and most valuable spherical or pear-shaped pearls. *Blister pearls* are found attached to the inside of the shell and are produced by a piece of irritant lying between the inside surface of the shell and the outside of the animal, which has been covered over by the pearly secretion.

Bivalve molluscs

The pearl-forming mollusc is a simple rudimentary creature consisting of a soft visceral mass enclosed within a double shell. The term 'bivalve' derives from the Latin 'bis' meaning 'two' and 'valvae' meaning 'leaves of a door'. Thus, the valves are the two halves of the shell. The animal itself more resembles a scallop than an edible oyster. Two large adductor muscles, one attached to each half of the shell, pull the shells together to close in the animal. The animal has a heart, mouth and alimentary system but no brain. On the ventral side of the adductor muscles are the gills. Near the mouth, adjacent to the hinge of the shells, is the foot. Near the base of the foot in a small pit is a gland that secretes a substance that forms a bundle of fibres, termed the *byssus*, which the animal uses to attach itself to surfaces on the bottom of the body of water. Of most importance is the *mantle* – a double fold of epithelial tissue above and beneath the animal that surrounds it loosely, but completely encloses it. The two halves of the mantle are joined along the hinge line of the shells. It is the mantle that is responsible for the formation of the shell.

When the adductor muscles relax it allows the shell to open, permitting seawater to enter. The water carries with it microscopic animal life which is the sustenance of the mollusc. The gills, which hang down into the mantle cavity, are responsible for oxygen absorption and also filter food particles from the water. Additionally, the wall of the mantle cavity contains capillaries and acts as a second respiratory surface.

On the outermost faces of the mantle is a layer of secretory cells, the ectoderm, which secretes the substance that forms the shell. The shell has three layers. The outside layer is a dark horny organic substance which approximates to the formula $C_{32}H_{48}O_{11}$ and which is termed *conchiolin*. This layer is known as the *periostracum*. The middle layer is composed of prismatic columns of minute crystalline calcium carbonate, $CaCO_3$, usually in the form of calcite. In this layer, the prismatic layer, the prisms are arranged at right angles to the surface of the shell and are held together by a 'cement' of conchiolin. Cells at the edge of the mantle secrete these two layers and, as such, once they are formed they cannot increase in thickness. The inner layer, which forms the internal surface of the shell, is secreted by the entire surface of the mantle. This is the nacreous layer that provides mother of pearl. It increases in thickness throughout the life of the animal. Nacre is built up of overlapping platelets of crystalline calcium carbonate in the form of aragonite; the principal crystal axes of these crystals are at right angles to the shell surface. This overlapping fashion is said to be reminiscent of roofing tiles, although in reality more closely resembles irregular ripples in the sand caused by gently lapping waves on a beach. The nacreous layer forms a smooth surface for the body of the animal to rest upon.

Pearl formation

When a shell is examined it is observed that the outside is rough and the inside is smooth to rest against the outer envelope of the animal. In early times, when microorganisms and parasites were not understood, the formation of pearls was explained as a mollusc's defence against mechanical intruders such as a grain of sand or sharp piece of shell. The mantle is connected to the edge of the shell all the way around; that is how the shell grows, and an inert bit of sand could have no route to get between the mantle and the shell. Thus, in the modern explanation, the intruders defended against are worm-like parasites – cestode, trematode or nematode – when their larvae are immobilized in certain parts of the mollusc. If the mollusc cannot expel the irritant, it is logical that it would attempt to cover it up with nacre, just as it did on the inside of its shell to produce a smooth surface, thereby easing the irritation. And, this is precisely what it does and how it forms a pearl. There are two basic means, each of which forms a different type of pearl.

One way is to secrete nacre over the irritant, cement it to the shell, and add layer by layer of smooth nacre on top. In this case, the bulge, or blister as it is called, produced on the shell may be removed and used as the jewel known as a *blister pearl*. Such pearls may be almost regular hemispheres or quite irregular in shape. They have a non-nacreous base, where they have been cut from the shell, and this can be hidden by the mount if desired. A different process produces true pearls, those whole pearls with a full nacre surround – that of encystation.

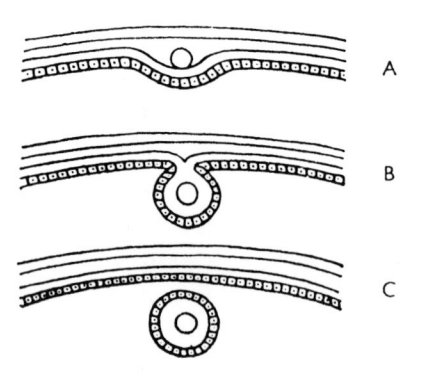

Figure 17.1 Illustration of the formation of a cyst pearl. A: a depression is formed in the mantle. B: the mantle depression deepens, trapping the irritant. C: the mantle depression joins together at the neck and produces a hollow sphere, the pearl sac, separate from the mantle tissue, leaving the pearl sac as a tumour or cyst within the body of the animal.

Encystation

This method involves the formation of a pearl inside the body of the animal, thus it produces the *cyst pearl*, or alternatively the *mantle pearl*. Gradually in successive stages a depression is formed in the mantle that traps the intruder. As the depression becomes deeper it forms a sac-shaped pouch. This eventually joins together at the neck producing a hollow sphere lined with epithelial cells quite separate from the mantle of which it was originally a part. This sphere of cells is termed the *pearl sac*. The wound in the mantle coalesces and leaves the pearl sac within the animal as a tumour or cyst. The nacre-secreting cells of the pearl sac are still living and go on secreting nacre over the irritant, thus building up the pearl. The nacre is the same as it is with shell production; however, in the case of pearls, there is generally no trace of the prismatic calcite layer and the whole of the pearl consists of nacre, the minute crystallites of aragonite formed in a framework of conchiolin. Further, these small crystals are deposited in layers, producing concentric layers like those of an onion, and have their vertical axes at right angles to the surface of these layers. Thus, the structure of a pearl is both concentric and radial, and it is upon this that the methods of testing pearls are based. Given a good position within the animal, the pearl is normally round in shape. If the position is unfavourable, it may produce baroque pearls, or drops or buttons.

The chemical composition of pearl is approximately 82–92% calcium carbonate ($CaCO_3$), most of which is in the form of aragonite; 4–14% conchiolin; 2–4% water and 0.4% other elements. Its structure is that of an aggregate composed of tiny orthorhombic aragonite crystals. Conchiolin is noncrystalline. The SG of pearl lies within the range of 2.60 to 2.78, largely dependent on region, and each region conforms to a narrower range of density. The non-nacreous pearls of the giant conch are over 2.8. The hardness of pearl is about 2.5 to 4.

The cause of the different overtone colours, such as the delicate rosée seen on fine pearls, is unknown. Similarly the cause of the body colour of nacreous pearls showing pronounced colour, such as golden-yellow, yellow, pink, blue, grey, gunmetal, bronze and black, is not clearly known either. It is believed to be related in some way to the colour of the shell and, to some degree, to the position of the pearl within the mollusc. The nature of the water has also been suggested as having an influence.

It may be that the beautiful spectral-coloured lustre of pearls is due to a combination of two optical phenomena – interference and diffraction – the combined effects of which are termed *orient*. Orient is seen on mother-of-pearl shell and on the finest pearls, but not on every pearl. It is an optical phenomenon in which spectral rainbow colours move across the surface in response to movement of either the object, the light or the observer. Traditionally orient has been explained as an interference phenomenon that is due to the thickness of the individual crystals of calcium carbonate, but only if the crystals were of appropriate thickness for thin-film interference to occur would it yield the iridescence. More recent research has brought to light the possibility that orient is in fact a diffraction phenomenon. Orient is not observed on every pearl because the microscopic crystals of calcium carbonate, which comprise the nacre, may not be of appropriate size, thickness and distribution pattern across the surface of the pearl to generate the optical effect on every pearl. When the platelets of aragonite are appropriately sized they produce a two-dimensional array that is capable of generating spectral colours by diffraction of light from the surface. How bright the generated colours appear depends upon the relative uniformity of the two-dimensional surface microstructure, which determines its effectiveness as a two-dimensional diffraction lattice. Thus, it may be that orient is due to the combined effect of interference of light at thin films and diffraction of light from the aragonite 'tiles' of the nacreous layer.

Table 17.1 Pearls by type, region and specific gravity

Location	Mollusc	Commonest colour	Specific gravity
Salt water pearls			
Persian Gulf	*Pinctada radiata*	Creamy-white	2.68–2.74
Gulf of Manaar	*Pinctada radiata*	Pale crème-white	2.68–2.74
North coast of Australia	*Pinctada margaritifera*	Silver-white	2.68–2.78
Northwest coast of Australia	*Pinctada maxima*	Silver-white	2.67–2.78
Shark Bay, western Australia	*Pinctada cacharium*	Yellow	2.67–2.78
Venezuela	*Pinctada radiata*	White	2.65–2.75
Japan (Akoya)	*Pinctada fucata martensii*	White, greenish tinge	2.66–2.76
Florida and Gulf of California	*Strombus gigas* (Queen conch)	Pink	2.81–2.87
Florida and Gulf of California	*Haliotis* (abalone)	Greens, yellows, blues, etc.	2.85
Gulf of California	*Pinctada margaritifera*	Black	2.61–2.69
Freshwater pearls			
North America	*Unio*	Often coloured	2.66–2.78+
Europe	*Unio margaritifera*	White	2.66–2.78+
Cultured pearls			
Japan	*Pinctada martensii*	White	2.72–2.78
Non-nucleated cultured pearls			
Japan	*Hyriopsis schlegeli*	White	2.67–2.70
Australia	*Pinctada margaritifera* or *Pinctada maxima*	White	Mean 2.70

Types of pearls

As previously noted, pearls may be natural or cultured. Traditionally, 'pearl' with no qualifier was used to denote only natural pearls; however, these are so rare today that the term has begun to be applied to the

Table 17.2 Types of pearls

Oriental	Natural, saltwater pearls from the Persian Gulf and the Gulf of Manaar. These are the finest pearls.
Akoya	Saltwater pearls from the Akoya oyster, *Pinctada fucata martensii* from Japan. Today most are cultured. They are roundish, typically light pink to white to yellowish, and range up to about 10mm.
South Sea	Cultured, saltwater pearls from the *Pinctada maxima* harvested in Indonesia, the Philippines, Australia, Thailand and Myanmar. White or golden-yellow colour and range in size from 9mm to 20mm.
Tahitian	Cultured, saltwater pearls from French Polynesia. Colour is naturally black, light to dark grey, greenish, bluish or brownish. Sizes range from 9mm to 20mm.
Biwa	Cultured, freshwater pearls from Lake Biwa in Japan.
Blister	Natural, saltwater pearls that grow attached to the shell on one side. When cut from the shell, this side is non-nacreous.
Mabe	Assembled cultured blister pearl. A half-bead nucleus is glued to the shell, covered with nacre, cut from the shell and the bead removed. The domed nacre is cleaned and filled with wax or another bead and sealed with a mother of pearl backing.
Seed	Natural, small pearls 2mm or less.
Keshi	General term for pearls that grow accidentally in a mollusc harvested for cultured pearls. Can be small like seed pearls or skinny and long up to 14mm. South Sea keshi can grow very large. Term originally used only for natural seed pearls found in harvested Akoya oysters.
Abalone	Saltwater pearls from animals from the *Haliotis* genus. Known for vivid colours: green, blue, yellow, pink, silver, etc. Often unique baroque shapes.
Conch	Saltwater pearls from the *Strombus gigas*, Queen conch. Pink, non-nacreous pearls with a porcelain-like surface.

cultured pearl. This is a misuse. The term 'pearl' without a qualifier should only denote natural pearls, all others being referred to as 'cultured pearls'. Another general categorization of pearl is 'saltwater', from the oceans or seas, or 'freshwater', from rivers and lakes. In general, saltwater pearls are markedly more valuable. Pearls may be of various types, as listed in Table 17.2. Other names are sometimes applied to pearl: *Button pearls* are cyst pearls which have rounded tops and flat bases resembling a button; *Drop pearls* are pear-shaped and are always cyst pearls; *Baroque pearls* are irregularly shaped pearls which may be cyst or blister in formation; and *Seed pearls* is the name applied to very small pearls.

Pearl-producing species

Species are categorized in descending order of specificity as: kingdom, phylum, class, order, family, genus and species. Animals from the phylum Mollusca, and the class Bivalvia, produce saltwater pearls. This is often cited as the class Lamellibranchia, and this is an alternative name. The order is *Pterioida*, and the family is *Pteriidaie*. Various species of the genus *Pinctada* produce commercial gem-quality pearls. These are all saltwater animals. *Pinctada radiata* is the most important: a small mollusc with a shell about 60mm in diameter. It was formerly known as *Pinctada vulgaris*, but has been renamed. It is fished for pearls only; its shell is too small for commercial harvest. It is fished from the Persian Gulf, the Gulf of Manaar, and historically the Red Sea. *Pinctada maxima* is fished in Australian waters. It is a larger mollusc known as the gold-lipped or silver-lipped oyster. The pearls are silver-white. This same colour is also produced by *Pinctada margaritifera,* the black-lipped oyster, fished off the north coast of Australia, and a green-lipped variety from the Gulf of California produces black pearls. Another small oyster, *Pinctada cacharium*, has a shell of 7.5cm to 10cm and produces yellow pearls at a fishery at Shark Bay, Australia. *Pinctada fucata martensii* produces the Japanese Akoya pearls.

Saltwater pearls are also known from animals of the genus *Strombus*, of the class *Gastropoda* and family *Strombidae*. This is a genus of medium to large sea snails, known as true conches. The *Strombus gigas*, the Queen conch or pink conch, produces a gem-quality pink pearl. These conch pearls are non-nacreous and have a porcelain surface characterized by 'flame-like' markings. Also from the *Gastropoda* class are those animals from the genus *Haliotis*, known as Abalone. These are univalves and produce highly coloured iridescent pearls that are often quite baroque.

Freshwater pearls are found in mussels of the genus *Unio*, which are medium-size bivalve molluscs in the family *Unionidae*, known as river mussels. Many pearls are also obtained from the niggerhead (*Quadrula ebena*), the bullhead (*Pleurobema oesopus*), the buckhorn (*Tritogonia verrucosa*) and some others.

Pearl-producing locations

The finest pearls, those termed *oriental pearls*, are fished exclusively from the Persian Gulf in an area that extends from Kuwait at the north to United Arab Emirates at the south; off the islands of Bahrain is the prime area. Oriental pearls are also fished from the Gulf of Manaar, a large shallow bay of the Indian Ocean between the southeastern tip of India and the west coast of Sri Lanka. Australia has multiple locations producing pearls, running from the west coast at Shark Bay up through the northwest coast and along the northern coast over the tip of the Cape York Peninsula of northern Queensland down to the fishing areas of the Coral Sea. South Sea fisheries are significant producing areas: the islands of Micronesia and Polynesia produce pearls that are often large and well shaped. The most important centre is Tahiti. Historically, Japanese fisheries cultivated pearls known as Akoya pearls, although the cultured pearl industry has rendered natural production virtually obsolete. Other areas include: off the coast of Venezuela and the Gulf of California; the Mergui Archipelago, off the coast of southern Myanmar; the Sulu Sea in the southwestern Philippines; around New Guinea and Borneo; and the Gulf of Mexico.

Freshwater pearls are found in the rivers of England, Scotland and Wales, and the rivers of North America as well as European rivers in Germany, Austria and Scandinavia.

Cultured pearls

It is now known that if a foreign body gets between the shell and the body of a suitable mollusc the animal takes steps to cover the intruder with nacre, thus producing a pearl. Experimentation began centuries ago into methods to induce pearl-forming animals to produce pearls. Knowing little of the science of the pearl-bearing mollusc, the Chinese in the thirteenth century found that if they inserted an object between the shell and the animal, it subsequently became coated with pearly nacre. Even to this day, metal figures of Buddha are so treated.

During the last decade of the nineteenth century the Japanese Kokichi Mikimoto commenced experimental production of blister pearls. He cemented mother-of-pearl pellets to the inner side of the shell of a mollusc. After the animal had been returned to the sea for some years and then fished up again, it was found that the pellet had been covered with nacre over the exposed surface. The bead was then broken away from the shell and the broken surface ground flat. A piece of mother of pearl was then pegged onto the base and ground to a symmetrical shape to produce the whole sphere. Alternatively, so-called *mabe pearls* are cultured blister pearls that have been grown on the shell over a soft talc-like bead. The original bead is removed, the inside of the hollow dome of nacre is cleaned and often tinted, and then a smaller bead is cemented into the cavity. It is then backed with a base of mother of pearl. If these pearls are mounted in a closed-back setting, the mother of pearl base is not seen but, if they are unset, the deceptive nature is at once apparent.

Although Mikimoto was the first to market fully spherical cultured pearls, he is not the inventor of the process. In all likelihood this is Tatsuhei Mise, who prior to 1904 produced a pearl in the species *Pinctada martensii* by means of a tissue graft around a tiny bead nucleus. In 1905 Tokichi Nishikawa repeated this method with gold and silver nuclei. They made a joint agreement for the Mise/Nishikawa method and a patent was granted to Nishikawa in 1916. By 1921 the cultured pearl, as we know it today, appeared on the market. The method employed to produce these pearls, highly technical in nature, has of recent years been brought to a mass-production basis.

In 1914 Mikimoto applied for a patent on a method invented by his good friend Otokichi Kuwabara. Their system, known as the 'all-lapped' system, involves fishing an oyster at maturity and inserting into its body, through an incision made with a scalpel, a mother-of-pearl bead in a sac made from the mantle cut from another mature mollusc that is killed by the operation. The sac containing the mother of pearl bead is wrapped and tied with silk thread and the wound in the second and living oyster is antiseptically treated. The oyster is then returned to the sea for a period of years, being kept in a wire cage. On being fished up after due time has passed, a period from three to six years, the bead is found to have been coated with a nacreous layer in exactly the same way as a mollusc produces a natural pearl. Today it is the Mise/Nishikawa method that is used for cultured pearls because the process of using a sac of mantle was found to be wasteful of molluscs – so now only a small patch of mantle about 2mm square is inserted into the oyster and this is followed by the bead. This small patch of epithelial tissue grows round the bead, forming the essential pearl sac. The Japanese cultured pearls are produced in the small oyster, *Pinctada martensii*, chiefly in Ago Bay near the Shima Hanto peninsula. Other bays on the eastern coastline of the main island of Honshu are also used for oyster farms. The controlled conditions keep the oysters in shallow waters that never fall below 10°C. The typical cultivation period is 3.5 years and oysters fished up in a shorter period – between twelve and eighteen months – will have thinner, less beautiful, less durable, nacre. Most of the cultured pearls are grown in inland waters of Japanese bays, but there are now fisheries in the Pacific Islands, in Australia, and other locations.

Non-nucleated cultured pearls

Since the end of the Second World War pearls have been cultivated without a mother-of-pearl nucleus. These pearls, which are rather characteristically bun-shaped and white in colour, are grown in a Japanese freshwater mussel known as *Hyriopsis schlegeli*. This is a large bivalve that abounds in Lake Biwa; it has internal nacre of fine colour. A graft of mantle tissue from another mussel is inserted into incisions made in the edge of the mantle of the live mussel. Up to twenty mantle grafts are inserted into each mussel. And of the 60% that produce pearls, nearly 100% of these yield the full twenty pearls. *Pinctada maxima* are used for non-nucleated cultured pearls in Australia. Clear detection of these pearls is not easy. X-ray shadowgraphs show either a large cavity or a fine patch that may not be at the centre of the pearl, and the Biwa pearls exhibit strong fluorescence under X-rays.

Detection of cultured pearls

Let us now consider the internal difference between a cultured pearl and a natural pearl, which from external examination appears much the same, and how this can be exploited for testing. The natural pearl, when sliced through, appears to be a series of concentric shells with, or without, a distinguishable nucleus. If the nucleus is present it is very small compared to that of the cultured pearl. On the other hand, a cross-section of a cultured pearl shows the parallel-banded structure of a large mother-of-pearl bead nucleus surrounded by a series of thin layers of true nacre, usually only 0.5mm to 1.0mm in total thickness. Hence, the parallel-banded bead shows directional properties. These are taken advantage of in testing with instruments.

As general observations, to the experienced eye, cultured pearls may show a greenish tinge and subcutaneous markings like 'varicose veins'. A string of cultured pearls 'twirled' before a desk lamp will often show telltale gleams as a result of reflections from the layers of the mother-of-pearl bead. If the light is transmitted through the bead, in certain directions where the layers are parallel to the light beam, light and dark stripes will be seen on the surface. Further, on looking down the drill canal it may be possible to see the junction layer between the bead nucleus and the outer nacreous shell.

As first placed upon the market as pearls from a new fishery, their true nature was not at once discovered – not until a pearl was broken and the mother-of-pearl bead discovered. Then came the necessity of finding a method whereby these pearls could be distinguished from the natural pearl, if the value of natural pearls was not to be adversely affected. Obviously, one could not break apart the pearls to determine if there was a bead core. The test had to be non-destructive. Initial tests were conducted on fluorescence under LWUV light. Cultured pearls were found to fluoresce with a greenish-yellow glow that was markedly different from the sky-blue fluorescence of natural oriental pearls. Unfortunately, it was discovered that green fluorescence was common to some natural pearls – in particular those from adjacent waters – and thus the method could not provide a definitive answer. About the same time, it was attempted to use the penetrating power of X-rays to see if the core would show up as a white shadow, similar to how bones appear on an X-ray in contrast to flesh. Owing to technical difficulties at the time, this technique was not pursued.

Examinations of specific gravity of natural and cultured pearls were conducted in 1922. It was found that if a suitable heavy liquid, bromoform diluted with benzene to where calcite suspends, SG 2.71, was used and pearls from a necklace were poured into the liquid, if the pearls were natural, about 70% floated and if they were cultured less than 10% floated. More recently, B. W. Anderson and C. J. Payne have shown that most cultured pearls have a relatively high specific gravity compared to the more restricted range of natural pearls. The higher SG is due to the large mother-of-pearl bead. They found that in general, with a liquid of SG 2.74, most cultured pearls sink and most natural pearls float. This provides a quick approximate test. Note that heavy liquids are damaging for organics such as pearls so the test, if undertaken at all, should be quick and the pearls cleaned thoroughly afterwards.

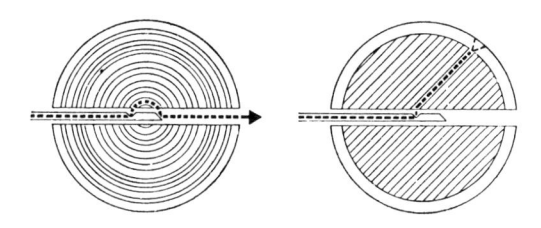

Figure 17.2 The endoscope. With the natural pearl on the left, when the endoscope is in the centre the upward beam of light enters one of the concentric layers and travels around it to be reflected from the second mirror and out along the drill canal where it is seen as a bright flash of light. With the cultured pearl on the right, the layered structure cannot return the light reflected from the first mirror. The reflected beam proceeds up and along the parallel layers of the mother-of-pearl bead to the surface where it may be seen as a streak of light on the outside of the pearl.

Nearly all pearls are fully drilled, meaning drilled straight through side to side, in order that they may be strung in necklaces, and this drill canal provides a means whereby the centre of the pearl may be examined. Probably the most satisfactory means of detecting drilled cultured pearls is by the *endoscope*. The apparatus consists of a strong source of light in a suitable housing that is directed down a fine hollow needle that has at its end two polished mirrors at angles of 45° in opposite senses. The first, interior, mirror causes the beam of light to reflect upwards and out of a small aperture cut in the top of the tube. The second, exterior, mirror forms the end of the tube. A small low-powered microscope views the surface of the second mirror. The optical system is arranged such that the operator can see the surface of the pearl when it is placed on the needle. When the needle is placed on the instrument and light is passed down the needle, a fine beam of bright light is seen directed up through the aperture in the top of the needle. If a natural pearl is placed on the needle, the upward beam of light enters one of the concentric layers and travels around it by a process of total reflection. This beam is 'lost' except when the needle is at the centre of the pearl when the light passing upwards from the first mirror is reflected along the adjacent concentric layer to be reflected from the second mirror and out along the drill canal where it is seen as a brilliant reflection in the microscope eyepiece. The pearl is passed back and forth along the needle and observed for the flash, which proves without doubt that the pearl is natural. Should the pearl be cultured, it has a layered structure and cannot return the light reflected from the first mirror to the end mirror, and therefore no flash is seen as the pearl is passed back and forth on the needle. In this case, the reflected beam will proceed up and along the parallel layers of the mother-of-pearl bead, and will be seen as a streak of light on the outside of the pearl. If the pearls are only half-drilled the method is not particularly useful except where a cultured pearl may show the streak of light on the surface even as the instrument penetrates the drill hole a short distance. This instrument requires some practice in operation, but in the hands of an experienced worker, upwards of 200 pearls an hour can be tested.

Should the pearl not be drilled, the endoscope method is not available, and another approach must be taken. In 1912, Max von Laue instigated practical experiments based on the theory that X-rays were electromagnetic radiation with a much shorter wavelength than visible light and that if a structure could be found in which regular divisions were close enough together, then if the rays were as suggested, they could be diffracted by such a structure in the same manner as light waves are diffracted by a finely ruled grating. Extension of his research led to obtaining diffraction patterns by passing a beam of fine X-rays through a crystal. The crystal's atomic lattice planes reflected the X-rays and formed *Laue spots* on the diffraction pattern. The pattern itself formed what is known as a *Lauegram*.

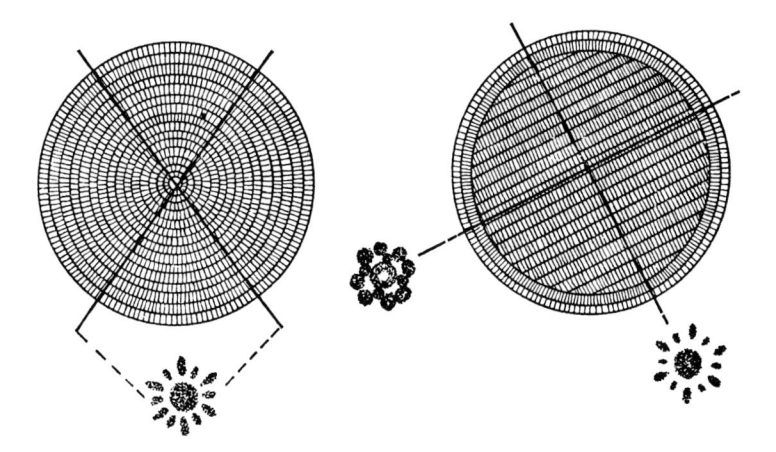

Figure 17.3 Lauegrams of natural and cultured pearls. The natural pearl on the left produces a lauegram having a hexagonal, six-spot, pattern in all directions. The cultured pearl on the right has one direction where it produces a six-spot figure; all other directions show four-spot figures.

Lauegrams of natural and cultured pearls should differ. In natural pearl, the aragonite crystallites, which have pseudo-hexagonal symmetry, are radially arranged. Thus, when a narrow beam of X-rays is passed through the centre of a natural pearl it must travel along the vertical axes of the aragonite crystallites regardless of the orientation of the pearl. This produces a lauegram having a hexagonal, six-spot, pattern. If, however, cultured pearl is examined, there is only one direction in which the beam will pass along the long axes, thus producing a six-spot figure; all other directions will cut across the short axes of the crystals and hence show four-spot figures. With a cultured pearl, the bead nucleus of mother of pearl has straight layers, and in one direction the aragonite crystallites are arranged perpendicularly. In this case, when a narrow beam of X-rays is passed through the pearl it passes at right angles to the layers and travels along the vertical axes of the crystallites and yields a hexagonal lauegram the same as that of a natural pearl. However, at right angles to this direction, the X-ray beam is travelling across the prisms of the crystallites and, as this is a direction of 4-fold symmetry, the lauegram produced is that of a four-spot pattern. In testing a pearl, should the first picture show a six-spot figure, the pearl must be rotated through a right angle, and a further test conducted. If this also shows a six-spot lauegram, the pearl is natural; if a four-spot figure is shown, the pearl is cultured.

The earlier failure to obtain a satisfactory X-ray picture of pearls has since been overcome. A change in technique now yields a *shadowgraph* or a *skiagram* that is of definite value in cultured and natural pearl differentiation. This is obtained by the use of fine-grain film that allows the structure of the pearls to be seen; and also by special techniques, such as immersing the pearls in liquid having similar X-ray density to the pearls, or by holding them in a wax such as 'plasticine'. The method is dependent upon the difference in X-ray transparency between conchiolin and calcium carbonate. In natural pearls, conchiolin may fill the centre of the pearl, as it does with 'blue pearls', or it may fill the layers between aragonite crystals. If these appear as circles or arcs near the centre of the pearl then it is natural. With cultured pearls, the mollusc does not like the job of coating the large irritant and often first coats the bead with a layer of conchiolin somewhat irregularly deposited. This appears as an outline around the bead nucleus, which shows up somewhat in shadow itself, against the transparent outer rim of nacre. If the bands of the mother of pearl bead are perpendicular to the surface of the film they may appear as weak stripes across the bead. Control of the film development is critical. The method will not always give a clear answer, but a great deal can be done in far less time than by the use of the lauegram or endoscope.

Fluorescence to ultraviolet radiation, whether LW or SW, is of no discriminatory value in differentiation of natural and cultured pearls; however, X-ray fluorescence is. Under X-rays natural pearls rarely glow but cultured pearls glow greenish-yellow. This glow is a weak response and may need to be viewed in a darkened room. The glow results from the mother-of-pearl bead that contains a trace of manganese. The test is not conclusive as all freshwater pearls and some saltwater pearls from Australia show a similar response.

Quality and price factors

Pearls of all types are judged and priced on a number of factors: (1) size – the larger the pearl, the rarer and more valuable it is; (2) lustre – the brilliance of the pearl. High-lustre pearls display strong bright light reflections; those of low lustre look dull and milky; (3) surface blemishes – the number and size of the blemishes. Some blemishes are normal and the smaller and less obvious they are the better the pearl;

(4) shape – in general, the more a pearl deviates from perfectly round the less valuable it is; (5) colour – white and pinkish saltwater pearls are most highly valued and pink overtones are most desirable. Natural blacks are also highly valued as long as they have overtones. White South Seas are the most prized, but golden-yellow can be just as valuable. With freshwater pearls the price is more variable with demand; and (6) nacre thickness – a judgement factor with cultured pearls only. The thicker the nacre the more valuable the pearl, as thick nacre produces both beauty and durability.

Owing to the organic constituent, pearls are not as durable as mineral gems. They are much softer. Moreover, pearls have a tendency to lose lustre and exhibit cracks as a result of the conchiolin drying out – such that too dry an atmosphere is detrimental to pearls in general – or acids dissolving the calcium carbonate. All that being said, pearls will hold their beauty and value for centuries if they are properly cared for. Acids in perspiration, the skin, hairspray and perfume are the typical culprits. Avoid direct contact with sprays and perfumes and clean pearls regularly. Acids from perspiration, cosmetics and perfumes that get on to the pearl string seep into the drill canal and attack the nacreous layers of the pearl. Pearl necklaces, as they are worn on the skin most often, should be cleaned and restrung every six months. Avoid storing pearls in cotton wool as it often contains acids.

Imitation pearls

The worldwide imitation pearl industry, which markets imitation pearl jewellery and necklaces from very low-priced to quite expensive, has its roots in Paris where in 1656 Jaquin produced the first imitation pearls. His imitations were hollow opalescent glass beads lined on the inside with a parchment size that adhered the *essence d'orient*, an iridescent composition made with scales from a fish, either the *bleak* or the *herring*, to the inside of the glass bead. The bead was then filled with wax to make it solid. Today it is common for imitation pearls to consist of a solid glass, plastic or mother-of-pearl bead upon which are placed several coats of *essence d'orient*. After each coat has been applied, it is burnished down. An imitation synthetic pearl essence is now often used for coating pearls. Some twenty coats may be applied to the better class of imitation.

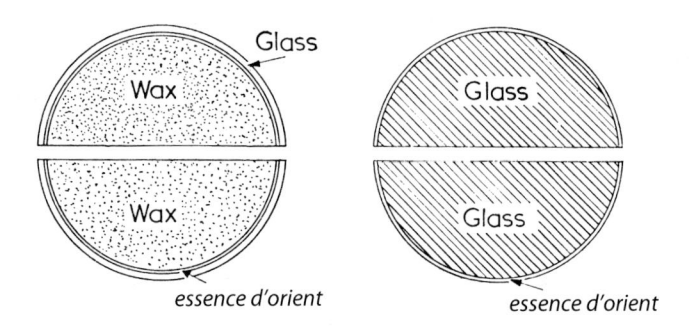

Figure 17. 4 Two types of imitation pearls. Left: hollow glass bead type. Right: solid glass bead type.

Identification is made with the loupe, as the imitation pearls do not have the fine overlapped rippled nature of the nacre of natural or cultured pearls. At the drill hole, the glass bead may show as a result of the wearing away of the *essence d'orient* from attrition against the adjacent bead. When rubbed over the teeth most imitation pearls feel smooth, and natural and cultured pearls feel gritty. Note that some recent imitations have a somewhat gritty feel. If a spot of ink is placed on the outside of an imitation with a hollow glass bead, the spot will show doubled due to the reflection from the internal surface of the glass sphere. Specific gravity is a good test as imitation pearls of the hollow bead type average about 1.55, which is much lower than that of natural and cultured pearls, and those with a solid glass bead are much higher in SG at 2.85–3.18. Polished spheres of hematite have imitated black pearls. These are easily detected by their much higher specific gravity, about 5.0, and by their red streak. Pink coral imitates pink pearl. It may be detected by the surface striations of the coral or the absence of the 'flame markings' typical for conch pearl. Pink pearl has a higher density: 2.85 to that of coral at 2.6–2.7. Any plastic imitation of pink pearl that lacks iridescence, as synthetic resins can imitate only this type, may be easily identified by the extremely low SG of the plastics.

Treatments

Natural pearls can be 'skinned', meaning that a poorly coloured or blemished layer is removed by careful abrasion. Cracks can be 'cured' by careful soaking in warm olive oil. The temperature must be strictly controlled because at about 150°C pearls turn brown and devalue. Unattractive dark-coloured pearls can be stained to an attractive black colour. This is done by soaking the pearls in a weak solution of silver nitrate and then exposing them to UV light that causes the silver to be reduced by the action of the organic constituent to a black powder. The pearl is polished to give lustre.

Cultured pearls can be stained rosé by bleaching in hydrogen peroxide and then soaking in a dye made of vegetable oil. Black colour is induced by silver nitrate as it is with natural pearls. Cultured pearls also turn black if exposed to gamma rays from a cobalt 60 nucleus for approximately sixteen hours. Some pearls have the bead nucleus coloured or dyed before nucleation. Surface-dyed pearls are detected by swabbing with a very weak solution of hydrochloric acid on cotton wool, which removes some of the dye, leaving a stain on the cotton wool. Bleaching is used to lighten or remove the undesirable greenish tint. It is not detectable and is accepted practice.

LESSON 18

Organic Gems

Amber

Despite the fact that it is a fossil, amber is a youngster compared with most gemstones, which typically formed when the Earth was a slowly cooling mass. Amber is organic, of vegetable origin. It is the fossil formed through natural polymerization of original organic compounds in the resin of ancient trees. This sticky resin seeped down the sides of trees and in geological time the trees were buried and the resin hardened. Amber has been found in sediments dating from 360 million years ago, although the average is 30–90 million years. Amber is a hydrocarbon with no set chemical formula as a result of its complex composition; the general formula is $C_{10}H_{16}O$. Many varieties of trees have produced resin that evolved into amber, including conifers from the order *Pinales*; and the families *Araucariaceae*, such as the monkey puzzle; *Taxodiaceae*, such as sequoia; *Taxaceae*, such as yew; *Pinaceae*, such as pine; and *Cupressaceae*, such as cedar and juniper. Amber was fashioned as early as 8000 to 6000 BC because of the ease in shaping it. It is cut into cabochons, beads, religious and other objects, and also left freeform.

Transparent to opaque, amber most characteristically is in tints of yellow and yellowish-red, although it may be found in shades of orange, brown, red, green, white, blue and black, as it tends to oxidize easily and darken with age. Amber is very soft, averaging only 2–2.5, about the same as a fingernail, although there are differences by location. Burmese amber from Myanmar is the hardest, ranging up to 3; Baltic amber is average, and Dominican amber may be as soft as 1–2. In general, hardness varies with age; the longer the amber has been buried the harder it becomes. Specific gravity lies between 1.04 and 1.10, averaging 1.08, making it just heavier than water and able to float on

ORGANIC GEMS

salt water. Clearer amber is heavier than amber with more bubbles. RI is 1.54. It is amorphous and tough to brittle, with conchoidal to splintery fracture and no cleavage. Lustre is resinous. It is likely to show anomalous extinction between crossed polars. It exhibits negative frictional electricity after being rubbed with a cloth and will pick up small pieces of tissue paper. In fact, this property is the origin of the English word 'electricity', as derived from the Greek word for amber, which is *elektron*. It is not a diagnostic property, as certain amber imitations also possess it; however, if there is no frictional electricity then it cannot be amber. Amber is a poor conductor of heat and feels warm to the touch. Amber exhibits fluorescence, typically greenish under SWUV, bluish-white under LWUV and less often green, white or orange, and this appears to be related to sulphur content. Amber softens at 180°C and melts at 290–300 °C.

Amber may be classified by transparency, country of origin, location such as land or sea, size and quality of the pieces, and the presence or absence of succinic acid. On the basis of transparency amber is described as: (1) clear – completely transparent, typically shades of yellow, the rarest; (2) massive – more cloudy in appearance; and (3) cloudy – ranging from translucent to opaque. Cloudy amber is further separated into: bastard – containing a vast number of bubbles; bony or osseous – white to brown opaque material having the general appearance of bone or ivory; fatty – with tiny bubbles and dust particles, usually translucent yellow resembling fat; and foamy or frothy – opaque, very soft, does not take a polish well and typically has pyrite infilling cracks.

Amber by country is as follows in order of importance: Baltic amber, *succinite*, from unconsolidated marine deposits near Gdansk in Poland and northeast of Kaliningrad, Russia, on the Samland peninsula. This *sea amber*, as it is called, is washed up on the shores by wave action from an outcrop of amber-bearing earth beneath the Baltic Sea, as well as dredged from shallow waters. It is also open-pit-mined, and in this case termed *pit amber*, from a rock stratum known as blue earth, a sedimentary clay, shale and sandstone mix 30–50m underground. It is washed in revolving cylinders with sand and water and graded into three categories: block amber – pieces large and clear enough to be used directly for fashioning into gems; pieces for pressed amber – pieces too small in themselves, but clean enough to be used for pressed amber; and all rough material which can be heated to obtain succinic acid, amber oil, and colophony resin used in making varnishes and lacquers. Other varieties have little, if any, succinic acid and are termed *retinite*. British amber is essentially the same material as Baltic amber as it was carried there by the tides of the Baltic Sea. The other significant commercial deposits are in the Dominican Republic. Burmese amber, *burmite*, is found in northern Myanmar. It is generally a pale yellow colour, usually not as clean as Baltic material, and calcite is often an inclusion. Sicilian amber, *simetite*, is of a darker colour, reddish hues being more common. Romanian amber, *rumanite*, is also deeper-coloured amber; brownish-yellow, brown, red and black are common colours. Amber also comes from Denmark, Sweden, Germany, Latvia, Malaysia, Mexico, Canada, the United States, and other countries, sometimes occurring in coal.

Amber inclusions are typically air-filled or water-filled bubbles. Rarer are plant inclusions such as flowers, seeds and pine needles; or mineral inclusions, such as pyrite. The most desired are insects and other small life forms. Insects are typically gnats, flies, bees and ants. Occasionally more exotic creatures are captured such as beetles, butterflies and spiders. The desirability and rarity of trapped fossils is a greater determinate of value than the colour and quality of the amber.

Amber is treated to improve clarity, colour and apparent value. Amber is often treated to shades deeper in colour in order to suggest more aged amber. Various shades of red and brown are common; green, as well as black, are colours sometimes produced. Clarification involves heating cloudy amber in an oil that boils at a temperature that will not cause decomposition or damage, such as rapeseed oil. It may or may not be stable and could possibly revert to cloudy after months or years. Another result of the boiling in oil treatment often seen is that internal stress cracks, termed spangles, appear. It is also possible to introduce dyes into these iridescent cracks, producing a variety of colours. In addition, because of the desirability and inherent value of amber with insect inclusions, these are often added to the amber after the fact. The amber is drilled out, the insect put in place, and the hole is plugged with a modern resin of a matching colour. Such insects are notably too perfect; a genuine insect inclusion will show signs of struggle and is often surrounded by gas bubbles due to decomposition. Furthermore, genuine insects are characteristically of extinct species. If the inclusion is of a species currently found in nature it is likely that it has been deliberately placed there, or the material is much younger copal, not amber.

Amber should be properly cared for as it is sensitive to hairspray and perfumes, which can cause the surface to dull. It should not be allowed to tumble around in a box with other jewellery as it is easily scratched by different stones and jewellery metals.

Amber has been imitated by: pressed amber, copal, resin, glass and various plastics. In pressed amber, or *ambroid* as it is known, smaller fragments of amber are heated to 200–250°C and pressed through a steel sieve to become amalgamated together to form a larger mass suitable for cutting. It may be readily detected by its fluidal structure, sharp margins of the zones of notably different colour or clarity, and elongated bubbles parallel to one direction. When bubbles appear in natural amber they are circular in form. It may also show disc-like stress fractures. It glows a brilliant chalky-blue colour under LWUV and this may make it easier to see its components. Be aware that copal is also pressed or reconstructed. Pressed amber, amber and copal all emit an odour of burning pine needles when a hot point is applied.

Copal is a more recent fossil resin from much younger trees, similar to true amber in colour and appearance. The SG is 1.08 and the RI is the same as amber. A quick test is to apply a drop of ether to the specimen in an inconspicuous place. Copal becomes sticky, leaving a dull spot on the surface. This is the result of its more recent hardening, whereas amber is unaffected. Copal has a tendency to craze at the surface, is much softer under the knife, and shows a whiter fluorescence under SWUV than amber. A New Zealand variety, better known as kauri gum, is sometimes classified as a copal but in fact it is even more recently formed and is properly classified as a resin rather than a copal on the resin–copal–amber continuum. It exhibits the same translucent pale-yellow colour as Baltic amber. It is more soluble in ether than amber, and also yields more readily to a knife.

Glass is much harder than amber and feels colder than amber. The SG is a great deal higher, lustre is vitreous and it does not peel or chip under the knife. Glass can usually be detected on sight and, if not, the saltwater test is an easy separation. Ten teaspoonfuls of salt in a glass of water is sufficient to produce a solution with an SG of about 1.13 in which amber, pressed amber and copal float, and glass and most plastics sink.

More convincing amber imitations are made of plastics. Celluloid, cellulose nitrate, has an SG of 1.38 to 1.42. It yields easily to a knife and shows yellowish-white fluorescence to UV and X-rays. Casein has an SG of 1.32 to 1.34, an RI of 1.55, shows a white glow under UV, and burns with the smell of burnt milk. Bakelite, a phenol formaldehyde condensation product, is the most used plastic imitation. Its SG is 1.26

to 1.28, and RI is 1.66. It is usually inert to UV, shows no frictional electricity, chars when burnt, and the smell is carbolic. Perspex is a polymerized acrylic ester with an SG of 1.18 and an RI of 1.50. It yields easily to the knife. Polystyrene has an SG of 1.05 and an RI of 1.59. It peels under the knife and has a much lower softening point than amber. The last two have not been used extensively, but can be convincing as a result of their SG. A hot needle brought near any plastic gives off the smell of burning plastic.

Jet

Organic of vegetable origin, jet is an opaque pure-black variety of fossil wood chemically related to brown coal, the carbonaceous fuel known as lignite. Its appearance is the origin of the adjective 'jet-black'. Jet beads have been found in burial places of civilizations dating back 10,000 years, and it was documented as an adornment as early as the fourteenth century by the poet Chaucer. Jet takes an excellent polish and was most popular in Victorian times in mourning jewellery. It remained popular through the flapper-era of the 1920s when it was worn in long strands, but has now fallen out of fashion. Jet results from high-pressure decomposition of wood from an ancient variety of conifer, *Araucaria araucana*, known as the monkey-puzzle tree. Hard jet is the product of carbon compression in which woody tissue is replaced by carbon and salt water. Soft jet is the product of carbon compression and fresh water. Only hard jet is suitable for jewellery. Its chemical composition is 75% carbon with smaller amounts of oxygen, hydrogen, sulphur and nitrogen. The most famous locality is Whitby in North Yorkshire, England. Other commercial production comes from Asturias, in Spain. It is also found in Germany, France and the United States.

Jet has a low hardness of 2.5 to 4, an SG between 1.30 and 1.40 with an average of 1.33, and an RI of between 1.64 and 1.68, typically with an indistinct edge around 1.66. Lustre is velvety to waxy and fracture is conchoidal. It is very brittle and inert to UV. It may be possible to see a wood structure under magnification of +120X. Jet is fashioned into beads, carvings and cameos as well as other objects. It can be engraved and used for inlay work.

It may be imitated by any of the plastics, particularly Bakelite and casein, the hard coal (anthracite) and also cannel coal, the hardened rubber (vulcanite), as well as glass and obsidian. Its superior polish can distinguish it from plastics. The SG of plastics may overlap the range of jet, but only in one case, that of Bakelite, does the RI of 1.66 also approximate. When a hot needle is applied, jet and cannel coal give off the smell of burning coal, vulcanite gives off the smell of burning rubber, Bakelite smells carbolic, and casein smells of burnt milk. Vulcanite shows mould marks and its black colour fades over time if exposed to sunlight. Jet is a poor conductor of heat and is warmer to the touch than glass. Obsidian has a higher SG of 2.33 to 2.42 and a single RI near 1.49. For a black material in jewellery, jet has been replaced in recent years by the harder and more durable stained black chalcedony. It is much heavier, SG 2.58 to 2.64, has an RI of 1.53–1.54, and it exhibits no streak – whereas the streak of jet is dark brown.

Coral

Coral is the internal axial skeleton at the base of and/or in the interior of a colony of small gelatinous coral polyps. The basal plate at the foot of a polyp attaches to a solid structure to hold the polyp in place. The calyx is a cup-shaped structure formed by a thickened annular calcareous ring atop the basal plate and into which the polyp can withdraw when threatened. The calices of adjacent polyps together become a skeletal platform holding the coral colony together. Subsequent generations attach their basal plates atop the earlier skeletal platform, thickening and enlarging it. It is the support structure, a type of scaffolding, for the soft animal bodies that sit on its surface. The chemical composition is almost entirely calcium carbonate ($CaCO_3$), together with about 3% magnesium carbonate ($MgCO_3$), and possibly a trace of iron oxide as well as 1.5–4% organic matter of indeterminate composition. The principal species used in jewellery are *Corallium rubrum* and *Corallium nobile*. These white to red corals occur in relatively small branching colonies attached to the sea floor as opposed to large coral reefs. The reddish hues are caused by the presence of organic pigments, known as transcarotenoides, in the organic matrix of the coral. Deep red and pink coral are the most valuable; angel skin coral is white to pale rose pink.

Coral was one of the first materials used for personal adornment. It is primarily fashioned into beads, cameos and small carvings. The finely ridged structure resulting from longitudinal stem canals connecting polyps becomes a finely corrugated structure running along the length of the branch. On a polished surface this is seen as the distinctive grained structure of coral: parallel stripes of slightly different colour or transparency. In cross-section it appears as a radial pattern.

SG is 2.6 to 2.7, hardness averages 3.5, and the RI is that of calcite: 1.49–1.65. It is polycrystalline. The lustre is dull when unworked, but that of polished coral is waxy. It is easily broken and the fracture is hackly. If sliced in a thin section it may be possible to see a spectrum with a band in the blue-green at 494nm, but it is not particularly helpful in testing. It glows weak pale violet to UV.

Black coral is the skeleton of *Antipathes spiralis*, which is found off the coast of Hawaii and off the Pacific coast of Queensland, in Australia. It is mostly organic conchiolin. SG averages 1.35 and RI is approximately 1.56. It is alternatively known as King's coral or Akabar. It shows concentric rings like a tree trunk structure. Lustre is resinous and fracture is splintery. Golden coral is light golden to dark yellowish-brown in colour, having a distinctive speckled appearance. It is mostly conchiolin, and SG averages 1.40 and RI is 1.58. It has a characteristic plucked chicken structure. Fracture is splintery and lustre is resinous. Blue coral from *Allopora subirolcea* is horny in nature and is known from the Cameroon coast.

Historically the Mediterranean Sea was the major source, but it is now known from various locations: the Atlantic off southern Europe and Africa, the Mediterranean Sea, the Red Sea, some parts of the Indian Ocean, and the Pacific off the coast of Japan and China. The best red coral is from off the coast of Algeria and Tunis.

Coral is stained to improve colour, most often to a deeper red, and this can be detected by acetone on a cotton swab. Black coral can be bleached with a 30% solution of hydrogen peroxide to give a golden coral colour. Under magnification the texture is different and it has slightly lower RI and SG. Much coral is reconstructed, a process in which small pieces are ground, plasticized and moulded into shape and repolished. The colour is very even as opposed to natural coral, which usually shows variation in depth of colour. Coral should be cared for properly as even mild acids readily attack its surface.

Gilson produces an imitation coral made from crushed calcite. Under magnification it appears quite homogenous, lacking the typical wood grain of coral. There may also show a concentration of colour in minute spots on the Gilson that is unseen in natural specimens. The Gilson is more porous and may show appreciable weight gain if soaked in water. The Gilson has an SG of 2.44, lower than any of the natural red or pink corals.

Ivory

Ivory is a general term for animal products of similar nature – specifically, dentine, the substance common to the teeth of all mammals as well as teeth modifications such as tusks. The chemical composition is calcium hydroxyphosphate with gelatinous organic matter.

Elephant ivory is primarily African elephant tusks. The finest material has a warm transparent mellow tint with little grain or mottling. Fossil ivory is obtained from the remains of the woolly mammoth, which lived about 200 000 years ago; however, it is not 'fossilized' because the tusks have not been altered by mineralization but it is a true ivory. Its noted location is Siberia where the tusks have been preserved in ice below the arctic ground. Hippopotamus ivory is from incisor and canine teeth. It is denser and of a finer grain than elephant ivory. Walrus ivory is canine teeth. It is less dense and coarser than that of the elephant and hippopotamus. Narwhal ivory is from a species of whale inhabiting the Arctic seas. It is an incisor tooth producing a coarse ivory. The boar and warthog supply coarse ivory more like the consistency of bone. It is suitable only for small articles.

The SG of elephant ivory varies from 1.7 to 1.9. Softer ivories have the lower values. Hippopotamus ivory is denser, about 1.90, and walrus and narwhal ivory averages 1.95. The RI in all cases is close to 1.54 and hardness averages 2.5. Ivory glows various shades of blue to UV. It is primarily carved into *objets d'art*.

Vegetable ivory is palm nuts, specifically Corozo nuts from the ivory palm. It is only found in small sizes and is pure white. Under magnification it shows a great number of torpedo-shaped cells running in roughly parallel lines. RI is 1.54 and SG averages 1.40.

Ivory tends to yellow over time due to atmospheric influences. It is common for bleaching to be used to lighten or remove staining. Some ivory is dyed to give the appearance of antique ivory; this is also common and undetectable.

Elephant ivory and mammoth ivory can be recognized in cross-section by patterns in different shades of crème colour. These are cross hatchings or engine turnings known as Schreger lines, and although both types of ivory have them, there is a difference in the angles formed by their intersections: those of elephant ivory have angles close to 120° and those of mammoth ivory have angles less than 90°. Many ivories show a grained appearance along the length of the tusk. A thin peeling of ivory immersed in a liquid of similar RI will show a wave-like pattern of fine fibrils under low-power magnification. Viewed end on, these appear as dots still in wave formation. In hippopotamus ivory the pattern is much finer and in walrus and narwhal ivory it is much coarser.

Bone has a different structure and a slightly higher SG of 2.00. Under the lens bone shows short dark lines or dots; canals that have been infiltrated with dirt. Deer horn resembles bone, but is characteristically browner in tint. SG is 1.70 to 1.85 and RI is approximately 1.56. Ivory is imitated by celluloid. A number of plastic sheets pressure-bonded together produce a grain-like effect. It is very sectile and easy to obtain a peeling that under magnification shows a fine-grained structure. The SG is also incorrect for ivory. Ivory is also often imitated by rhinoceros horn, which is tightly packed horny fibrils, not dentine. Its SG is 1.3.

Odontolite

This is a rarely seen fossil ivory sometimes known as bone turquoise or fossil turquoise. It is teeth from the mastodon and other extinct animals, from beds in the south of France. The ivory has become blue because of the presence of the ferrous phosphate mineral vivianite. As mined, it is a dull grey, but after heating it becomes a fine turquoise blue colour. It has an SG just over 3.00 and an RI of 1.57 to 1.63. Hardness is about 5. The striped appearance of ivory is evident under the microscope. It contains calcium carbonate, which will effervesce vigorously when in contact with hydrochloric acid.

Shell

Shell is the calcareous armour of shellfish, particularly those with an iridescent lustre such as the large pearl oysters *Pinctada maxima* and *Pinctada margaritifera*. Known as mother-of-pearl, it is ornamental and used for inlay as well as knife handles, buttons and other decorative objects. Some dark-coloured shell from the black-lip oyster suitably cut produces a cat's eye effect. It is popular for buttons and may be mounted in jewellery. Other varieties include: paua – brightly coloured blue and green nacre from the genus *Haliotis* found in Queensland, California and Florida; helmet shell – used for cameos that stand out in white bas-relief against a brown background; and giant or queen conch – used for cameos in which the carving is white on a rose-coloured background, known from the West Indies. Mother-of-pearl is often stained various colours by soaking in organic dyes. The colour is relatively unstable.

Tortoiseshell

This is a material obtained from the shield of a sea turtle. It is similar to horny tissues that make up horns, claws and nails and consists of a protein known as keratin, which has a complex composition. The most highly prized is that with rich-brown mottling on a warm translucent yellow background. It is obtained from turtles that inhabit the Maluku Islands in Indonesia. SG is typically 1.29, hardness is 2.5, and RI is 1.55. It is readily sectile. The clearer yellow portions glow bluish-white to UV. It is used for various articles including decorative hair combs and small boxes.

Under the microscope, the mottling of tortoiseshell is seen to be made up of spherical spots of colour; the closer together the spots, the deeper the colour. In imitations such as plastics, the colour appears in patches or swathes and there is no dot-like or disc-like structure seen under magnification. Imitations glow more variably to UV. Chips of tortoiseshell fuse to a black mass and smell of burning hair when a hot point is applied.

LESSON 19

Secondary Gem Species

This lesson provides brief notes on gem species that are encountered in the trade with lesser regularity, including faceted, ornamental and collector stones. It is necessary to learn the appearance and major properties of these gemstones.

Alabaster

Hydrous calcium sulphate, $CaSO_4 \cdot 2H_2O$. Gemmologically significant variety of gypsum. Monoclinic and produces excellent crystals that are often twinned in swallowtail forms; this variety is known as *selenite*. Alabaster is a massive variety. SG 2.30 to 2.33. H 2. RI 1.52 to 1.53. BI 0.009 to 0.010 is not usually detectable. Purest form is white and translucent. Varies from translucent to nearly opaque: a trace of ferric oxide produces light-brown and orange-coloured bands and veins and other impurities produce yellow, browns and black in veins and patches. One direction of perfect cleavage with two directions of distinct cleavage on single crystals. Finely granular fracture. Glows brownish to UV. Found embedded in limestone in Canada, Tuscany in Italy, and also England.

Andalusite

Aluminum silicate, Al_2SiO_5. Orthorhombic of prismatic habit with vertically striated prisms of nearly square cross-section. SG 3.15 to 3.17. H 7.5. RI 1.63–1.64, with BI of 0.007 to 0.013. Biaxial negative. Dispersion is 0.016. Transparent: greenish-brown, yellowish-green to rich green. Striking trichroism visible face-up as brownish-pink, yellowish and brownish-green. Note that it is incorrect to misrepresent the pleochroism as 'alexandrite-like', as this confuses pleochroism with colour change. Deep-green variety shows manganese spectrum: graded absorption band ending at 553.5nm, fine lines at 550.5nm, 547.5nm, 518nm and 495nm, with strong absorption of blue and violet. May be iron bands at 455nm and 436nm. Inert under LW, but brownish-green stones from Brazil glow dark green to yellow-green under SW. From Sri Lanka and Brazil. Inclusions are apatite prisms, rutile needles either single or in groups resembling a sea urchin and various curved inclusions.

Apatite

Calcium phosphate with some fluorine or chlorine, $Ca(F,Cl)Ca_4(PO_4)_3$. Hexagonal with stumpy prismatic habit, sometimes tabular. H 5. SG 3.17 to 3.23. RI 1.63 to 1.64, with BI 0.003. Uniaxial negative. Transparent: colourless, yellow from Mexico, blue and green from Sri Lanka and Myanmar may show chatoyancy, intense blue from Brazil, green from Ontario in Canada and violet from Czech Republic. Blue has strong dichroism: blue and pale yellow. Lustre is vitreous and dispersion 0.013. Yellow and blue show didymium spectrum. Yellow: a group of 7 lines around 580nm and a group of 5 lines around 520nm. Blue: broader bands, the strongest at 512nm, 491nm and 464nm. Fracture is conchoidal to uneven and cleavage is imperfect, leaving a rough surface in two directions. Yellow is lilac to UV; blue is bright blue to sky blue; green is mustard-yellow; and violet is greenish-yellow to LW and pale mauve to SW.

Benitoite

Barium titanium silicate, $BaTiSi_3O_9$. Hexagonal with tabular habit. SG 3.65 to 3.68. H 6 to 6.5. RI 1.757–1.804, with BI of 0.047. Uniaxial positive. Dispersion is 0.044. Transparent: blue often zoned with lighter blue from San Benito in California; pink is rare; and colourless. Blue has strong dichroism: blue and colourless. Blue glows bright chalky blue to SW and inert to LW. Colourless glows dull red to LW.

Brazilianite

Hydrous sodium aluminium phosphate, $Al_3Na(PO_4)_2(OH)_4$. Monoclinic with prismatic habit. SG 2.980 to 2.995. H 5.5. RI 1.603–1.623, with BI of 0.020. Biaxial positive. Dispersion is 0.014. Transparent: yellow to yellow-green. Discovered in Minas Gerais in Brazil. Cleavage is perfect in one direction parallel to one of the pinacoid faces. Fracture is conchoidal and it is brittle. Inert to UV with no definite spectrum. Dichroism is weak: slight change in shade of same hue.

Calcite

Calcium carbonate, $CaCO_3$. Trigonal with rhombohedral or prismatic habit. SG 2.71. H 3. RI 1.486–1.658, with BI of 0.172. Uniaxial negative. Dispersion is 0.017 for the ordinary ray and 0.008 for the extraordinary ray. Transparent to opaque: pure is colourless, also white, grey, red, pink, brown and yellow. Iceland spar is colourless. Satin spar is a fibrous aggregate that is often chatoyant. Cleavage is perfect in three directions. Dichroism is weak to none. Fluorescence is variable.

Danburite

Calcium borosilicate, $CaB_2(SiO_4)_2$. Orthorhombic with prismatic habit. Transparent yellow, yellowish-brown, brown and colourless. Light pink is rare. SG 3.00. H 7. RI 1.630–1.636, with BI of 0.006. Biaxial positive for yellow sodium light. As the angle between the optic axes is nearly 90°, the optic sign is negative for red to green light and positive for yellow to violet light. Shows a weak didymium spectrum. Exhibits a sky blue glow to UV and phosphoresces reddish when heated. Named after place of first discovery – Danbury in Connecticut. Known from Myanmar, Madagascar and Japan.

Diopside

Calcium magnesium silicate, $CaMg(SiO_3)_2$. Monoclinic with prismatic habit. SG 3.29. H 5.5 to 6. RI 1.675–1.701, with BI of 0.026. Biaxial positive but usually shows a spot reading of 1.68. Transparent to opaque: typically bottle green, also colourless, brownish and rarely violet-blue. Fibrous produces chatoyancy and asterism. Chromium-rich stones are brighter green and iron-rich stones may be so dark as to appear black. India is noted for a dark, almost black, variety that exhibits a four-rayed star. Cleavage is distinct prismatic in two directions. Lustre is vitreous. Some greens glow green to LW and are inert to SW. Chrome diopside shows spectra with: two sharp lines at 508nm and 505nm, a band at 490nm, fuzzy bands at 635nm, 650nm and 670nm and a strong doublet at 690nm. Duller greens show weaker spectrum: ill-defined bands at 505nm, 493nm and 446nm. Chrome diopside is from South Africa and Myanmar, which is also noted for cat's eyes. Diopside is also found in Sri Lanka, Brazil, Canada and the United States. A typically massive translucent to opaque dark blue-violet variety is known as *violane* and used for beads, carvings and inlay.

Enstatite

Magnesium iron silicate, $(Mg,Fe)SiO_3$. Orthorhombic. SG 3.26 to 3.28. H 5.5. RI 1.663–1.673, with BI of 0.010. Biaxial positive. Transparent: green and brownish-green. Green colour is found with diamond in South Africa and brownish-green from Myanmar. Orange-brown, yellow-brown and green are known from Tanzania. Rarely colourless, which has slightly lower constants. May show chatoyancy and very rarely asterism. Spectrum shows a characteristic sharp clear line at 506nm and also a line at 550nm. Cleavage is prismatic in two directions. Dichroism is weak: green and yellowish-green. Inert to UV. As iron content increases, enstatite merges into opaque hypersthene. An iron-rich variety with a bronze-like lustre is known as *bronzite* and is found in Austria.

Fluorite

Calcium fluoride, CaF_2. Isometric commonly with cubic habit and, more rarely, as octahedra. Cubes are often bevelled at the edges and the faces frequently show very low pyramids that produce striations on the surface. Interpentrant twinning is common. SG 3.18. H 4. RI 1.434. Dispersion is 0.007. Transparent to translucent: violet, blue, green, brown, yellow, pink and colourless. Blue John or Derbyshire spar is massive banded patterned with lines of blue, violet and purple and used ornamentally. Perfect octahedral cleavage. Fracture is conchoidal, but in compact types it may be splintery. Strong bright sky blue to violet is common under LW, a green glow is rarer, and Blue John is inert. SW response is weaker. Spectrum is vague but green shows weak bands at: 634nm, 610nm, 582nm, 445nm and a strong band at 427nm. Also known as fluorspar. Finest is found in England, also Switzerland, the United States and Canada.

Hematite

Compact type of iron oxide, Fe_2O_3. Trigonal found in large nodular masses called kidney ore due to its shape. Only occasionally found as rhombohedral crystals with iridescent tarnished faces. SG 4.9 to 5.20. H 6.5. RI is academic 2.94–3.22, with BI of 0.28. Uniaxial negative. Opaque gunmetal dark grey to blue-black. Metallic lustre. Splintery or granular fracture. Usually attracted by magnet. Often cut as beads, cabochons, buff-top cabochons, or carved as seal stones, intaglios and cameos. Leaves red on a streak plate. From Isle of Elba, Switzerland and England.

Idocrase

Complex calcium aluminium silicate, $Ca_{10}Mg_2Al_4(SiO_4)_5(Si_2O_7)_2(OH)_4$, with some isomorphous replacement of aluminium by iron and calcium by magnesium. Tetragonal with square prismatic habit. Prisms are often capped with pyramids terminated with the basal pinacoid. Transparent to opaque yellowish-green, green, yellowish-brown and light blue, grey and white. These varieties are alternatively known as *vesuvianite*. Massive green variety is *californite*. Blue variety containing copper is known as *cyprine*. SG 3.32 to 3.47 but massive green is between 3.25 and 3.32. H 6.5. RI 1.712–1.721, with BI averaging 0.005–0.012. Uniaxial with optic sign changing from negative with lower refractive indices to positive with higher refractive indices. Dispersion is 0.019. Vitreous lustre. Spectrum shows a strong band at 461nm and a weak band at 528.5nm. Brown stones sometimes show rare earth bands 574.5nm and 591nm. Inert to UV. Found in Italy, Siberia, Switzerland and Canada.

Iolite

Complex silicate of magnesium and aluminium, $Mg_2Al_4Si_5O_{18}$, with replacement of magnesium by ferrous iron and manganese and replacement of aluminium by ferric iron. Orthorhombic with short pseudo-hexagonal prismatic twin habit. SG 2.57 to 2.66. H 7 to 7.5. RI 1.53–1.55, with BI of 0.008 to 0.012. Biaxial negative. Dispersion is 0.017. Transparent to translucent: sapphire blue, rarely colourless. Very strong eye-visible trichroism: dark violet-blue, light blue and yellow. The spectrum shows bands at: 645nm, 593nm, 585nm, 535nm, 492nm, 456nm, 436nm and 426nm. In the violet-blue direction the 645nm and 426nm bands are masked by cut-offs at the ends of the spectrum. Inert to UV. Found in gem gravels in India, Sri Lanka, Myanmar and Madagascar. Sri Lankan stones can show hexagonal platelets of goethite or hematite, usually in parallel orientation which, if profuse, can give the stone a red colour when viewed broadside on. Mineral species name is cordierite.

Lapis lazuli

A rock that is a complex aggregate of several minerals: primarily: lazurite, calcite and pyrite. May also contain hauyne and sodalite as lazurite is an isomorphous combination of hauyne and sodalite. All four minerals are isometric and belong to a group of rock-forming minerals known as feldspathoids, which are formed when the silica content of the rock is insufficient to form true feldspar. Opaque to slightly translucent blue to greenish-blue to rich purple-blue. Iron pyrites may be present as brassy specks. SG 2.7 to 2.9. H 5.5. RI about 1.50. Shows spots and streaks of orange or copper colour to UV, which is weaker under SW. Formed by metamorphic action on impure limestone through contact with intrusive igneous rocks. Famous location is Afghanistan and other sources are Chile, California and Myanmar. Gilson manufactures a ceramic lapis lazuli simulant. It is available with or without pyrite inclusions. SG 2.46, H 3 and RI 1.53 to 1.55. The Gilson lapis is composed of mainly synthetic ultramarine and two hydrous zinc phosphates. These constituents are both absent from the natural lapis, and therefore it is a simulant – not a synthetic. Gilson-created lapis shows a uniformity of grain size and colour under magnification as well as a regular distribution of crushed pyrite fragments, whereas in natural lapis lazuli the pyrite is irregular crystalline grains.

Malachite

Hydrated copper carbonate, $Cu_2(OH)_2CO_3$. Monoclinic with botryoidal masses of radiating microcrystals as habit. SG 3.8. H 3.5 to 4. RI 1.655–1.909, but only a mean value of 1.86 is obtainable, with academic BI of 0.25. Biaxial negative. Opaque green typically with concentric bands of dark green and paler green. Inert to UV. Brittle and breaks easily. Streak is light green. Ornamental cut as cabochons, beads, *objets d'art* or used for inlay. Formed through the dissolving of copper ores and later deposition in rock cavities and veins. Effervesces when touched with acid. Found historically in Russia, and currently in the United States and Australia.

Moldavite

Natural glass composed of 75% silica, 10% alumina and smaller quantities of oxides of iron, potassium, calcium, sodium, magnesium, titanium and manganese. Isotropic. SG 2.34 to 2.39. H 5.5. RI 1.48 to 1.50. Transparent green, brownish-green or brown. May contain profuse bubbles. Spectrum shows two ill-defined lines in the blue and a vague band in the orange. Inert to UV. Most is from the Czech Republic. It is a tektite, arguably formed by the impact of large meteorites on the Earth's surface.

Obsidian

Natural glass composed of up to 77% silica and 10–18% alumina. Isotropic. SG 2.33 to 2.44. H 5 to 5.5. RI 1.48 to 1.51, usually near 1.49. Fracture is conchoidal. Lustre is vitreous. Usually opaque black or grey, but may be yellow, red, green, blue or brown. Formed by the rapid cooling of molten lava. Under magnification, it contains many small crystallites due to incipient crystallization. Networks of curving cracks may be characteristic. Widespread occurrence, but most is from North America.

Pyrite

Iron sulphide, FeS_2. Isotropic with habit of striated cubes or pyritohedrons having 12 five-sided faces. Opaque brass-yellow with metallic lustre. SG 4.95 to 5.10. H 6 to 6.5. Worldwide occurrence. Good crystals known from the Isle of Elba. Greenish-black or brownish-black streak. Pyrite is known as fool's gold.

Rhodochrosite

Manganese carbonate, $MnCO_3$. Trigonal with usually massive habit. Rhombohedral crystals are rare. SG 3.50 to 3.65. H 4. RI 1.600–1.820, with BI of 0.22. Uniaxial negative. Only the clear nearly transparent material shows this RI and BI; most shows a diffuse shadow edge near 1.71. Translucent to opaque lighter rose red with concentric banding of different shades of pink. Spectrum shows a band at 551nm and other vague lines, but it does not assist in identification. Some shows a dull red glow to UV. Effervesces when a drop of acid is applied. Found in Argentina, the United States and Romania. Rare transparent red stones are known from Namibia.

Rhodonite

Manganese silicate, $MnSiO_3$, with some calcium usually present. Triclinic and usually massive. Translucent to opaque: rose red or red and pink with black veins of other manganese minerals. Transparent crystals are rare. SG 3.6 to 3.7. H 6. RI averages 1.73. Transparent stones have shown RI of 1.733–1.747, and BI 0.014. Biaxial positive. Spectrum shows a broad band at 548nm and narrower lines at 503nm and 455nm. Ornamental and cut as cabochons. Cleavage is perfect in two directions but obscured by the aggregate structure. It is brittle. Lustre is vitreous to pearly on fractured surfaces. Inert to UV. Transparent shows dichroism: orange-red and brownish-red. Known from Russia, Australia and the United States.

Scapolite

Silicate of aluminium with calcium and sodium. Part of an isomorphous series between sodium-end member marialite, $Na_4Cl(Al_3Si_9O_{24})$, and calcium-end member meionite, $Ca_4(SO_4CO_3)(Al_6Si_6O_{24})$. Tetragonal with prismatic habit. Transparent colourless. Transparent to translucent pink and violet that are distinctly fibrous. Opaque yellow. Some stones are chatoyant. SG 2.60 to 2.74. H 6. RI 1.54 to 1.58, with BI of 0.009 to 0.020. Uniaxial negative. Dispersion is 0.017. Yellow stones are at the higher range of RI and BI. Perfect cleavage in two directions. Lustre is vitreous. Dichroism is strong in violet stones: dark blue and lavender blue. Weak dichroism in yellow stones: colourless or pale yellow and yellow. May exhibit yellow-orange fluorescence to LWUV and pink to SWUV. Yellow stones may show lilac or mauve under UV. Spectrum of pink and violet stones shows bands at 663nm and 652nm, with a strong absorption of the yellow. Usually found in metamorphic rocks from Brazil, Madagascar and Myanmar.

Serpentine

Hydrated silicate of magnesium, $H_4Mg_3Si_2O_9$. Monoclinic and habit is always massive. Varieties: purer translucent to oil green colour is *williamsite*, known as precious serpentine; resinous waxy yellow to yellow-green is *retinalite*, known as common serpentine; pale translucent light green to yellowish-green is *bowenite*, it often contains whitish cloudy patches; and rock masses with other minerals is *serpentine marble*. Bowenite is SG 2.58 to 2.59. H 5. RI averages 1.56. Spectrum shows bands at 497nm and 464nm that are not diagnostically significant. Inert to UV. Williamsite RI averages 1.57. H is 2.5. SG near 2.61. Weak whitish-green to LW, but inert to SW. Serpentine marble varies in SG from 2.5 to 2.7. H 2.5. RI averages 1.56. Serpentine fracture is granular to uneven. It is carved and used ornamentally. Groups or chains of greenish flakes of chlorite are often enclosed in bowenite. Williamsite may contain black octahedral inclusions. Produced by a number of geological processes, including the alteration of basic rocks of igneous origin or of metamorphic pyroxenes. From the South Island of New Zealand, Pennsylvania, South Africa and India.

Sinhalite

Magnesium aluminium iron borate, $Mg(Al,Fe)BO_4$. Orthorhombic, almost always found as rolled pebbles, rarely as good crystals. Transparent pale yellow-brown to golden or greenish-brown to black. Pink is rare. Darker colours attributed to increased iron content. SG 3.47 to 3.48. H 6.5. RI 1.67–1.71, with BI of 0.038. Biaxial negative. Dispersion is 0.017. Trichroism is distinct: light brown, greenish-brown and dark brown. Spectrum shows four bands in the blue and blue green at 493nm, 475nm, 463nm and 452nm and general absorption of the violet. Inert to UV. From Sri Lanka.

Soapstone

Acid metasilicate of magnesium, $Mg_3Si_4O_{10}(OH)_2$. Massive variety of the mineral talc or *steatite*. Monoclinic with massive habit. Sub-translucent to opaque white to silver-white; impurities may impart green, brown, yellow or red colours. It is often mottled or veined. Talc has hardness of 1 but, owing to impurities, soapstone may be up to 2. SG 2.2 to 2.8. RI 1.54–1.59, with BI of 0.05. Biaxial negative. Only a vague edge is seen at 1.54 for the massive variety. Ornamental stone used almost exclusively for carvings. Inert to UV. Often stained to look like jade. Lustre is waxy and feel is greasy. Widespread in northern Canada, and also found in central Africa and Zimbabwe.

Sodalite

Sodium aluminium silicate with sodium chloride, $Na_8Al_6Si_6O_{24}Cl_2$. Isometric and usually of massive habit, very rarely as dodecahedra. Opaque blue, often with white veining. It may also be green, yellowish, grey, white or light red, but only the blue is used gemmologically. SG 2.15 to 2.35. H 5.5 to 6. RI 1.48. Exhibits patchy orange fluorescence to LWUV. Cut as cabochons, beads and used for inlay. Notably found in Ontario and British Columbia, Canada and also Massachusetts, Norway, India and Brazil. It is one of the components of lapis lazuli, but the colour of sodalite is more royal blue and less saturated than that of fine lapis.

Sphene

Silicate of titanium and calcium, $CaTiSiO_5$. Monoclinic with wedge-like crystal habit; often twinned. Transparent yellow, brown or green, typically brownish-green or yellowish-green. Black and reddish-brown varieties are alternatively known as *titanite*. SG 3.52 to 3.54. H 5.5. RI 1.885 to 2.050, with BI of 0.105 to 0.134. Biaxial positive. Dispersion is 0.051. Lustre is sub-adamantine to resinous, cleavage is distinct in two directions and it is brittle. Weak rare earth spectrum. Trichroism is strong: brownish-yellow, reddish-yellow and light yellow to nearly colourless. Inert to UV. Brilliant cut and mixed cut to maximize fire. Found in cavities in gneiss and granite, and also in metamorphic schists and certain granular limestones in Switzerland, Canada and Madagascar.

Verdite

Massive ornamental muscovite rock coloured deep green by chromiferous mica fuchsite. Often containing red or yellow spots. SG 2.80 to 2.99. H 3. RI 1.58. Spectrum shows three lines in the deep red and a vague line in the blue. Inert to UV. From South Africa, Swaziland and Vermont.

Zoisite

Silicate of calcium and aluminium, $Ca_2(Al.OH)Al_2(SiO_4)_3$. Orthorhombic. Transparent blue-violet coloured by vanadium is known as *tanzanite*, also green, yellow, pink, brown and massive green. Massive pink is known as *thulite*. Rarely chatoyant. SG 3.35. H 6 to 7. RI 1.691–1.700, with BI of 0.009. Biaxial positive. Massive, may be SG 3.10, RI 1.70

and H 6. Inert under UV. One direction of perfect cleavage. Pronounced trichroism in tanzanite: sapphire blue, purple-red and greenish-yellow. Absorption spectrum shows a broad band in the yellow-green near 595nm with two fainter bands at 528nm and 455nm, with several weak lines in the red. Heat treatment produces tanzanite colours from some brownish material. It is permanent and undetectable. Famous location is Tanzania.

Index

Abalone pearl, 290
Aberration (lens), 160, 163
Absorption spectrum, 135–141
　See also Gemstones
Achroite, 186
Achromatic (lens), 160
Acicular, 44, 177
Adamantine, 52, 87, 171, 219
Adularescence, 88, 92–93
Agate, 44, 90, 195–196, 252
Alabaster, 312
Alexandrite, 130, 190–192
　Fluorescence, 144, 148, 156
　Spectrum, 191
　Synthetic, 192, 215, 222
Allochromatic, 129–130
Alluvial deposits, 14, 170
Almandine garnet, 157, 183–184, 245, 247, 280
　Spectrum, 139
Alpha (RI), 109, 111–114
Alpha particles, 23, 149, 189, 257
Amazonite feldspar, 203
Amber, 153, 302–306
Ambroid, 305
Amethyst, 193–194
　Synthetic, 226
　Treatment, 251, 253, 256
Ametrine quartz, 194
Amorphous, 31
Andalusite, 12, 131, 312–313
Anderson, B.W., 116, 136, 144, 213, 295
Andradite garnet, 183, 185–186

Angel skin coral, 307
Angstroms, 75
Anion, 23–25, 28
Anisotropic, 36, 105–107, 124
Annealing, 257
Anomalous double refraction (ADR), 126
Anomalous extinction (AEX), 126–127, 171
Apatite, 27, 55, 107, 313
　Fluorescence, 148, 156
　Spectrum, 140
Aplanatic (lens), 160
Aquamarine, 134, 176–178
　Simulants, 183, 213, 236, 248
　Spectrum, 177
　Treatment, 251, 256
Aragonite, 91, 285–288, 297–298
Archimedes' Principle, 61–62
Argon, 18–19, 29, 232
Asbestos, 89
Asterism, 79, 88, 90–91, 272
　Treatment, 174, 255–256
Atomic mass unit, 18
Atomic number, 17–21
Atomic structure, 12, 32, 48, 153
Atomic weight, 19–21, 60, 155
Atoms, 15–17, 22–25
Autoclave, 224–225
Autoradiograph, 259
Aventurescence, 88, 90, 92
Aventurine, 92, 94, 194–195
Aventurine feldspar, 92, 204
Aventurine glass, 94

Axes,
   Crystallographic, 35–42
   Optic, 36, 107–108
Axes of symmetry, 34–35

Baddeleyite, 217
Baguette cut, 279, 281
Bakelite, 305–306
Balance (weighing), 62–64
Baroque pearls, 287, 289–291
Basal cleavage, 55–56, 179
Basal pinacoid, 42
Bead (cut style), 279–280
Becke method (RI), 118–119
Bellataire, 261
Benitoite, 150, 313
Benzene, 67, 295
Beryl, 176–178, 260
   Aquamarine, 134, 176–178
   Bixbite, 176–177
   Emerald, 176–178
   Goshenite, 176, 178
   Heliodor, 176, 178
   Luminescence, 150
   Maxixe, 176, 178, 260
   Morganite, 176–178, 256
Beryl glass, 236
Beta (RI), 109, 111–114
Biaxial, 36, 40–42, 56, 104–114
   And dichroscope, 132
   And polariscope, 125–126
Bipyramid, 37–44
Birefringence, 36, 97, 100, 104–114
Biwa pearl, 289, 293
Bixbite, 176–177
Black opal, 91, 197–198, 244, 252–253
Blister pearl, 284–285, 289, 292
Block amber, 303
Bloodstone, 196
Blue John, 150, 316
Boart, 173
Bonding (chemical), 22–27, 54, 171
Bone, 196, 303, 309–310

Bone turquoise, 310
Boron, 19, 29, 171–172, 237
Botryoidal, 44, 196, 199, 318
Boule, 209–214
Bowenite, 320–321
Boyle, R., 226
Brachy axis, 40, 42, 188
Brazilianite, 314
Brewster angle, 115–117, 123
Brewster angle meter, 115–117
Bridgman, P., 228
Brightness, 128
Brilliance, 93–94, 96, 100, 173
Brilliant cut, 86, 271, 276, 281
Brillianteerer, 270–271
Briolettes, 279–280
Brittleness, 57
Bromoform, 67–68, 118
Bruting, 267–271, 275
Bubbles, 150, 165–167, 207, 244, 248
   Air, 63
   In amber, 303
   In glass, 188, 200, 239–242
   Gas, 192, 207, 211–219
Bytownite, 204

Cabochon cut, 89, 126, 279–280
Calcite, 27, 39, 156, 314, 317
   And dichroscope, 131–132
   And polarized light, 121–123
Cameo, 272, 281
Canada balsam, 123, 254
Cape Series, 141, 148–149, 171, 231, 260
Carat weight, 173, 281
Carbonado, 173
Carbonates, 27
Carborundum, 231, 271
Carnelian, 195
Casein, 305–307
Cathay stone, 241
Cathodoluminescence, 143, 149, 231
Cation, 23–25

INDEX

Cat's eye, 79, 89, 191–192, 194, 241, 272, 280
   Chrysoberyl, 89, 191–192
   Quartz, 89, 194
Catseyte, 241
Celluloid, 305, 310
Centipede inclusion, 167
Centre of symmetry, 33–34, 39, 42–43
Chalcedony, 32, 57, 86–87, 89, 125, 194–195
   Treatments, 252
Charles & Colvard Ltd, 232
Chatham synthetic emerald, 221
Chatoyancy, 79, 88–89
Chelsea Filter, 121, 133–134
Chemical element, 16–19, 26, 29, 129–130
Chemical vapour deposition (CVD), 231
Chemiluminescence, 143
Chloromelanite, 201–202
Chromatic aberration (lens), 160, 163
Chrome tourmaline, 186
Chromium, 19, 28–29, 129–130, 133–134, 144, 176
   Spectrum, 139
Chrysoberyl, 89, 190–192
   Fluorescence, 144, 148, 156, 191–192
   Spectrum, 139, 191
Chrysoprase, 195
Church, A.H., 135
Citrine, 193–194, 251, 256
Clarity grading, 165, 173, 251
Cleavage, 31, 36, 47–48, 50–51, 53–56
Cleaving diamond, 267–269
Clerici solution, 67
Closed form, 43
Cobalt, 20, 28–29, 134, 139, 147, 157
Cold light, 142

Cold melt, 217
Colour, 127–128
Colour centres, 129, 179, 193
Colour change gemstones, 130, 141, 174, 185, 190–192, 210
Colour zoning, 165, 167, 175, 226, 230, 255–259, 272–273
Complementary colours, 127–129
Composite gemstones, 243–250
Composite light, 128
Compounds, 12, 15, 17, 22, 26–28
Conch pearl, 287–288, 290, 300, 311
Conchiolin, 283–287, 298–299, 308
Conchoidal fracture, 57–58, 239–240
Conductivity,
   Electrical, 151, 171, 173, 258–259
   Photo, 151
   Thermal, 153–154, 173, 232, 240
Conoscope, 125
Contact liquid, 102
Contact rotation twin, 45
Contrast immersion (RI), 119
Coober Pedy, 197
Copal, 70, 304–305
Copper, 20, 27–29, 130, 153
Coral, 159, 205, 253, 307–309
Cordierite, 317
Core (Earth), 10–11
Cork, J.M., 258
Cornelian, 195
Corozo nut, 309
Corundum, 29, 55–57, 89, 154, 173–175
   Cause of colour, 129
   Hardness, 48–53
   Luminescence, 150
   Synthetic, 181, 207–211, 215, 222–223
   Treatments, 254–257, 263–264
Cosmic radiation, 75
Covalent bonding, 24
Crackle dyed quartz, 253

Critical angle, 94, 98–102, 107, 123, 276
Crocidolite, 89, 196
Crookes, Sir W., 228, 257–258
Cross cut, 281
Crossed filters, 143–144
Crossed polars, 124–126
Crown glass, 135–136, 235–238
Crust (Earth), 10–11
Cryptocrystalline, 32, 194–196, 199
Crystal(s), 31–33
Crystal classes, 33
Crystal forms, 42–44
Crystal habit, 43–44
Crystal systems, 36–42
Crystalline materials, 31
Crystallographic axes, 35–42
Cubic crystal system, 37–38
Cubic zirconia, 217–219
Culet, 268, 270–271, 273–275, 277
Cultured pearls, 291–298
Cushion cut, 277–278, 281
Cut styles, 273–282
Cutting, 267–273
CVD, 231
Cyclotron, 258
Cymophane, 89, 190–191
Cyst pearl, 286
CZ (cubic zirconia), 217–219
Czochralski pulling, 215–217

D1 sodium line, 75
Danburite, 314
Dark field illumination, 160–164
Demantoid garnet, 167, 185–186
  Spectrum, 140
Dendritic quartz, 193
Density,
  Optical, 79, 97–98
  Physical, 59–72
Deuterium, 18, 23–24
Deuterons, 23–24, 258
Diamond, 151–154, 157, 170–173
  Cleavage, 54–55
  Cutting, 267–271
  Styles of, 273–282
  Doublets, 244
  Fancy colour, 172
  Formation, 170
  Fracture filling, 172, 262–264
  Grading, 173
  HPHT, 172, 260–262
  Irradiation, 257–260
  Laser drilling, 172, 264–266
  Luminescence, 149
  Properties, 171–173
  Spectrum, 141, 171
  Synthetic, 173, 226–231
  Treatments, 172, 253
Diaphaneity, 86
Diasterism, 90, 194
Diatomic, 26
Dichroic, 132
Dichroism, 131–133
Dichroscope, 131–133
Didymium, 140, 236, 313–314
Differential selective absorption, 131
Diffraction, 85, 90–91
Diffusion treatment, 256–257
Di-iodomethane, 67–68
Diopside, 89–90, 314–315
Dioptase, 41
Direct measurement (RI), 115–116
Directional hardness, 51–52, 172
Disclosure of treatments, 251, 263, 266
Dispersion, 82–84, 94–95
Distant vision method of RI, 114–115
Dodecahedron, 37–38, 171
Dome form, 42
Double cut, 274
Double refraction, 105
Double refraction divergence, 106–114
Doublets, 243–250
Dravite, 186
Durability, 13, 47
Dyeing of gemstones, 252–253

INDEX

Earth, 9–11
  Geologic layers of, 10–11
Ebonite (vulcanite), 307
Einstein's quantum theory, 74
Electrical conductivity, 151, 171, 173, 258–259
Electroluminescence, 143, 149
Electromagnetic spectrum, 74–77
Electromagnetic waves, 74
Electron, 16–18, 22–28
Electron bombardment, 258
Element (chemical), 15, 17–22
Elements of symmetry, 33–35
Elephant ivory, 309–310
Eluvial deposits, 14
Emerald, 176–178
  Fluorescence, 148, 150
  Inclusions, 167, 176–177
  Spectrum, 140, 177
  Synthetics, 68, 219–225
  Treatments, 178, 254
Emerald cut, 269, 280–281
Emerald filter, 133–134
Emission spectrum, 135
Encystation (pearl), 286
Endoscope, 295–297
Energy levels, 22, 28, 145
Enhancements, 251–266
Enstatite, 315
Epiasterism, 90
Epigenetic inclusions, 167
Epsilon (extraordinary ray), 108
Essence d'orient, 299–300
Etch pits (trigons), 165, 171
Ether, 73
Euhedral, 43, 183, 186, 188
Eureka can, 65
Excited state (electron), 146
Extinction position, 124
Extraordinary ray, 108

Facet condition, 50, 164
Faceting, 52, 267–281
False beta, 113–114

False cleavage, 56
Fancy coloured diamonds, 172, 253, 262, 281
Fancy shapes, 277–282
Faraday, M., 74
Feathers, 86, 167, 175, 177, 191–193, 216, 223
Feil, C., 207
Feldspars, 202–204
Fibre optic light, 138
Filters,
  Chelsea, 133–134
  Crossed, 143–144
  Polarizing, 123, 168
Fire, 83, 94–96, 197
Fire opal, 197
Flame fusion, 207–215
Flint glass, 235–238
Fluorescence, 142–150
Fluorite, 150, 315–316
Flux melt, 222–223
Flux reaction, 221
Flux transport, 220
Foiling, 252–253
Form (crystal), 35–44
Fossils, 11
Four C's (diamond grading), 173
Fracture, 57–58
Fracture filling, 262–264
Fraunhofer lines, 83, 135
Fremy, E., 207
Frequency, 77
Freshwater pearls, 288–289, 291, 293, 298–299

Gadolinium gallium garnet (GGG), 216–217, 222
Gahnospinel, 182
Gamma (RI), 109–114
Gamma radiation, 75, 259
Garnet, 89–90, 125, 183–186
Garnet-structured simulants, 215–217
Garnet-topped doublet, 245–247

Gaudin, M., 207
Gay-Lussac, J., 207
Gemstones, 10–14
   Cardinal virtues, 13
   Classifications, 13–14
   Definition, 12–13
   Deposit types, 14
   Formation environments of, 10–12
General Electric (GE), 228–230, 260
Geneva ruby, 207
German lapis, 252
GGG, 216–217, 222
Giant conch, 287
Gilson created opal, 198
Gilson emerald, 220
Gilson imitation coral, 308–309
Gilson simulated turquoise, 199
Girdle, 276
Glass, 125, 234–242, 305
   Cobalt spectrum, 139
Globular form, 44
Goldstone, 241
Goshenite, 176, 178
Grain (diamond), 268
Granite, 10, 12
Granular habit, 44
Graphite, 17, 32–33, 48, 172–173
Grossular garnet, 183, 185
Ground state (electron), 145–146

Habit (crystal), 42–44
Hackly (fracture), 57
Hall, H. Tracy, 228
Hand lens (loupe), 159–160
Hannay, J.B., 227
Hardness, 47–53
   Directional, 51–51
   Mohs scale, 48–49, 53
   Testing, 49–52
Heat treatment, 254–260
Heavy liquids, 66–71
Heliodor, 176, 178
Hematite, 242, 316

Hemihedral, 43
Hemimorphic, 43
Herbert Smith refractometer, 101
Hessonite garnet, 185
Hexagonal crystal system, 37–39
Hiddenite, 200
Hippopotamus ivory, 309–310
Hodgkinson, A., 157
Holohedral, 43
Horsetail inclusion, 167
HPHT, 260–262
Hue, 128
Huygens, C., 73–74
Hydrogrossular garnet, 185
Hydrophobic, 198
Hydrostatic weighing, 62–64
Hydrothermal growth, 223
Hydrothermal reaction, 223–225
Hydrothermal transport, 225–226

Iceland spar, 123
Ideal brilliant cut, 276–277
Identification of gemstones
   (see individual species)
Idiochromatic, 129–130
Idocrase, 316–317
Igneous rock, 10–12
Imitation gemstones, 205, 234–250
Immersion cells, 166–167
Immersion technique (RI), 117–119
Impregnation, 253–254
Incident illumination, 164
Inclusions, 197
   See also Gemstones
Indicolite, 186
Industrial diamonds, 173
Infrared radiation, 75
Inorganic gemstones, 9, 12–13
Intaglio, 272, 281
Interference figure, 125–126
Interference of light waves, 84–85, 90
Interpenetrant twin, 45
Iolite, 317

INDEX

Ionic bonding, 24–25
Ions, 22–25
Iridescence, 88, 90
Iron, 157
Irradiation treatment, 257–260
Isometric crystal system, 37–38
Isomorphous replacement, 28–30, 179, 183, 186, 188, 202
Isotropic, 79, 105–106, 124
Ivory, 309–210

Jadeite, 201–202, 252–253
Jades, 201–202
Jasper, 196
Jet, 306–307

Kauri gum, 305
Kimberlite, 170, 226
KM treatment, 265–266
Konoscope, 125
Kunzite, 150, 200
Kyanite, 41, 51

Labradorescence, 90
Labradorite, 90, 203–204
Lamellar twinning, 45–46, 90
Lamproite, 170
Lamprophyres, 170
Lapidary, 13, 267, 272
Lapis lazuli, 242, 252–254, 317–318
Laser drilling, 264–266
Lattice diffusion, 257
Laue, M., 296
Laue diffraction pattern, 155
Lauegram, 296
Lavoisier, A.L., 226
Lead glass, 235–238
Life, 94, 96
Light, 73–96
    Waves, 74–77
Light field illumination, 160–164
Lily pad inclusion, 188
Limestone, 11
Lithification, 11

Long wave UV light, 147–150
Loupe, 159–160
Luminescence, 142–152
    Mechanism, 145–147
    Testing, 147–152
Lustre, 87–88
LWUV, 147

Mabe pearl, 289, 292
Macle, 45, 171, 278
Macro axis, 40, 188
Magma, 10
Magnetism, 157–158
Malachite, 318
Malaia garnet, 185
Maltese cross, 125
Mammoth ivory, 309–310
Manganese, 28–29, 130, 147
Mantle (Earth), 10–11
Mantle (oyster), 285–286, 292–293
Marble, 11
Marcasite, 87
Marquise cut, 277, 279
Massive crystalline, 32
Maxixe beryl, 176, 178, 260
Maxwell, J.C., 74
Melange, 282
Melanite, 186
Melee, 281–282
Metallic lustre, 87
Metamers, 128
Metamict, 32
Metamorphic rock, 11–12
Metasomatism, 12
Metastable level, 146
Methylene iodide, 67–68
Metric carat, 173
Microcline feldspar, 202–203
Microcrystalline, 194, 196
Microscope, 159, 162–169
Mie scattering, 93
Mikimoto, K., 292
Mineral, 12–13
Mixed cut, 281

Moh's scale, 48–49, 53
Moissan, F.F.H., 227, 232
Moissanite, 154, 231–233
Moldavite, 318
Molecules, 26
Monobromonaphthalene, 67–68
Monochroic, 132
Monochromatic light, 103, 128
Monoclinic crystal system, 37, 40–41
Moonstone, 92–93, 167, 203
Morganite, 176–178, 256
Morse, H., 275
Moss agate, 196
Mother-of-pearl, 310–311
Mtorolite, 195
Myrickite, 195

Nacre, 90, 285–294, 298–300, 311
Nanometer, 75
Narwhal ivory, 309–310
Nassau, Kurt, 207
Negative reading (RI), 103
Nephrite, 201–202
Neutron, 16–18, 22, 24, 259–260
Neutron bombardment, 259
Newton, Sir Issac, 74, 127–128, 226
Nickel, 157
Nicol prism, 121–123, 168
Noble, Sir A., 228
Noble gases, 18, 22–23, 25, 29
Non-crystalline, 31
Non-nucleated pearls, 288, 293
Non-spectral colours, 128
Normal (optics), 78
Nucleus (atomic), 16–17, 22, 24, 145–146

Objective lens, 163
Obsidian, 10, 318–319
Octahedral cleavage, 54–55, 316
Octahedron, 37–38, 55, 171, 181
Ocular lens, 163
Odontolite, 310

Oiling of gemstones, 175, 178, 195, 199, 254
Old European cut, 275
Old mine cut, 275
Oligoclase feldspar, 203–204
Omega (ordinary ray), 108
Onyx, 196
Opal, 91, 196–198, 252, 254
    Composite stones, 244–245, 249–250
Opal essence, 241
Opalescence, 93
Opaque, 86
Open form(s), 43
Optic axis, 36, 106–114, 126
Optic character, 100, 105, 119
Optic sign, 108
Optical density, 79, 97–98
Ordinary ray, 108
Organ pipe spectrum, 144
Organic gemstones, 9, 302–311
Orient (pearl), 88, 90–91, 287
Ortho axis, 41
Orthoclase feldspar, 202–203
Orthorhombic crystal system, 37, 39

Padparadscha sapphire, 210, 257, 260
Painting gemstones, 253
Paraiba tourmaline, 186
Parallel growth, 46
Paris, L., 212
Particle theory, 74
Parti-coloured tourmaline, 186
Parting, 56
Paste, 234–242
Pavilion, 274–278
Pearls, 260, 283–301
    Cultured, 291–298
    Formation, 285–288
    Imitation, 299–300
    Price factors, 298–299
    Treatment, 300–307
    Types, 288–290

INDEX

Pearly lustre, 87–88
Pedion, 42
Pegasus Overseas Limited (POL), 260
Pegmatite, 10, 12
Peridot, 106, 188–189
    Spectrum, 140
Peristerite, 204
Peruzzi cut, 275
Phenakite, 41, 192, 223, 225
Phosphorescence, 143–149
Photoconductivity, 151
Photoluminescence, 143
Photomicrography, 168
Photon, 74
Piezoelectric effect, 152
Pinacoid, 42
Plagioclase feldspar, 202–204
Planck, M., 74
Plane polarized light, 106, 131–132
Plane of symmetry, 33–34
Plasma, 196
Plastics, 242, 305–307
Play of colour, 88, 91
Pleochrosim, 130–133, 167
Point cut, 273
POL, 260
Polariscope, 124–127, 168
Polarized light, 121–127, 131–133
Polarizing filters, 123, 168
Polaroid, 123
Polycrystalline, 124, 168
Polymorph, 32
Polysynthetic twinning, 45
Potch opal, 197, 244
Prase, 195
Prasiolite, 194
Precious stones, 14
Pressed amber, 303, 305
Primary deposits, 14, 170
Prism (crystal form), 42
Protogenetic inclusions, 167
Protons, 16–18, 22–24, 258
Pseudomorph, 44, 196

Pycnometer, 66
Pyralspite garnet series, 183–185
Pyramid (crystal form), 42
Pyrite, 319
Pyroelectricity, 152–153
Pyrope garnet, 183–184
Pyrophyllite, 228–229

Quantum theory, 145–146
Quartz, 126, 152, 192–196
    Cat's eye, 89
    Crystalline, 192–194
    Interference figure, 125–126
    Multi-crystalline, 194–196
        Chalcedony, 195
        Jasper, 196
        Pseudomorphs, 196
        Quartzite, 194–195
    Refractometer example, 109–111
    Synthetic, 222–223
    Treatment, 253–257, 260
Quartzite, 194–195

Radiant cut, 278
Radioactive isotope bombardment, 257–258
Radioactivity, 258–260
Radium treatment, 257–258
Rainbow quartz, 193
Rare earth elements, 138, 140, 216, 219, 241, 317, 322
Rarity, 13
Ray(s) (concept), 76
Read, Peter G., 116
Reconstitution method, 230
Reconstructed amber, 305
Reconstructed coral, 308
Reconstructed ruby, 207
Reflection of light, 78–79
    Laws of, 78
    Total internal reflection, 95, 99–102, 123
Refraction, 79–82
    Laws of, 81

Refractive index, 79, 82, 97–120
Refractometer, 98, 100–120
Refringence, 79, 97
Resinous lustre, 87
Retinite, 304
Rhodochrosite, 319
Rhodolite garnet, 184
Rhodonite, 319–320
Rhombic dodecahedron, 37–38
Rock, 10–13
   Definition, 12–13
   Types, 10–12
Rock crystal, 193
Rondisting, 270
Rontgen, W., 154
Rose cut, 278
Rose quartz, 90, 194
Round brilliant cut, 275–277
Rubellite, 186
Ruby, 89, 173–175, 254
   Heat treatment, 174, 254–257
   Inclusions, 175
   Luminescence, 150, 174
   Simulants, 234–241, 243–250
   Spectrum, 139, 174
   Strontium titanate, 214–215
   Synthetic, 207–212, 222–223
Rutilated quartz, 193
Rutile, 26, 41, 83
   Needles, 184, 210, 255, 265, 313
   Synthetic, 213–214

Sagenitic quartz, 193
Salt solution (for amber), 70
Sandstone, 11
Sapphire, 89, 173–175, 254, 260
   Heat treatment, 174–175, 254–257
   Inclusions, 175
   Luminescence, 150, 174
   Simulants, 234–241, 243–250
   Spectrum, 139, 174
   Synthetic, 207–212, 222–223
   Synthetic spectrum, 140–141

Sard, 195
Sardonyx, 196
Saturation, 128
Sawing diamond, 269
Scaife, 270
Scapolite, 320
Scenic agate, 196
Scheelite, 41, 55, 148, 156
Schiller, 88
Schorl, 186
Scintillation, 95–96
Scratch test, 49–50, 52
Sea amber, 303
Secondary deposits, 14, 170
Sedimentary rock, 11–12
Selective absorption, 129–141
Self-contained melt, 217
Semi-precious stones, 14
Serpentine, 320–321
Sheelite, 150
Sheen, 79, 88, 92–93
Shell, 310–311
Short wave UV light, 147–150
Siberite, 186
Silk inclusions, 175
Silky lustre, 87–88
Simple form, 43
Simulants, 205, 234–250
Single cut, 274
Single refraction, 105
Sinhalite, 321
Skull melt, 206, 217–219
Slocum stone, 241
Smith, Dr G.F. Herbert, 101
Smoky quartz, 193
Snell's laws, 81
Soapstone, 321
Sodalite, 321–322
Sodium light, 103
Soudé stones, 243–250
Specific gravity, 59–72
   Liquids, 66–71
Spectra, 138–141
   See also Gemstones

Spectrophotometer, 260
Spectroscope, 134–141
  Technique, 136–138
Spectrum, 127
Spessartite garnet, 183–184
Sphene, 322
Spherical aberration (lens), 160
Spinel, 89, 181–183
  Luminescence, 150, 182
  Spectrum, 140, 182
  Synthetic, 182, 212–213
  Synthetic spectrum, 139, 213
Splintery (fracture), 57
Spodumene, 200
Spot RI method, 114–115
Staining of gemstones, 252–253, 308
Staligmitic, 44
Star stones, 89–90
Steatite, 321
Step cuts, 279–281
Stokes, G., 142–143
Striations, 46, 165, 174, 216, 300, 315
Strontium titanate, 214–215
Sub-adamantine lustre, 87
Subconchoidal fracture, 57
Succinite, 303
Sunstone, 92, 204
Surface diffusion, 256–257
Swiss lapis, 252
SWUV, 147–150
Symmetry, 33–35
  Elements of, 33–35
Syngenetic inclusions, 167
Synthesis techniques, 205–233
  Czochralski pulling, 215–217
  Flux melt, 222
  Flux reaction, 221
  Flux transport, 220
  High-pressure solution, 226–231
  High-pressure sublimation, 231
  Hydrothermal growth, 223
  Hydrothermal reaction, 223–225
  Hydrothermal transport, 225–226

  Skull melt, 217–219
  Verneuil flame fusion, 207–215
Synthetic gemstones, 205–233
  Corundum, 207–212
  Cubic zirconia, 217–219
  Diamond, 226–231
  Emerald, 219–226
  Moissanite, 231–233
  Rutile, 213–214
  Spinel, 212–213
  Quartz, 222–223, 225–226

Tabby extinction, 126, 213
Table cut, 273–274
Table of elements, 19–21
Table facet, 276
Tablet cut, 273
Tabular habit, 44, 174, 186, 313
Talc, 49, 321
Tanzanite, 322
Tausonite, 214
Tennant, S., 226
Tetragonal crystal system, 37–38
Tetrahedral bonding, 171
Thermal conductivity, 153–154, 173, 232, 240
Thermoluminescence, 143, 146, 149
Thetis hair stone, 193
Thorium, 32, 189
Three-phase inclusions, 167
Thulite (zoisite), 322
Tiger's eye, 89, 196
Titanium, 28–29, 129–130, 209–210, 213–217, 236–237
Tolkowsky, M., 276–277
Tone (colour), 128
Tongs, 161
Topaz, 30, 152, 178–181
  Treatment, 254–257, 260
Tortoiseshell, 311
Total internal reflection, 95, 99–102, 123
Tourmaline, 152, 186–188, 260
  Refractometer example, 108–111

Transition elements, 27–30, 130
Translucent, 86
Transparency, 86
Trap cut, 280–281
Treatments, 251–266
    Diamonds, 251–253, 257–266
    Disclosure of, 251, 263, 266
    Fracture filling, 262–264
    Heat, 254–260
    HPHT, 260–262
    Laser drilling, 264–266
    Pearl, 300–301
Triboelectric effect, 153–154
Triboluminescence, 143, 149
Trichroic, 132
Triclinic crystal system, 37, 42
Tricone burner, 209, 214
Trigonal crystal system, 37, 39
Trigons, 165, 171
Trillings, 190
Triple cut, 274–275
Triplets, 243–250
Tsavolite (garnet), 185
Turquoise, 198–200, 242, 254
Tweezers (tongs), 161
Twinned crystals, 44–46
Two-phase inclusions, 177–178, 187, 191, 193
Tyndall scattering, 93
Type I and II diamond, 171

Ugrandite garnet series, 183, 185–186
Ultraviolet light, 75
Umbrella effect, 258
Uniaxial, 106–111
Unit cell, 33
Uranium, 32, 138, 175, 189–190, 195, 237, 240

Uvarovite garnet, 185

Valence, 22–29
Vanadium, 28–29, 130, 144, 176, 185–186, 210, 217
Vegetable ivory, 309
Verdelite, 186
Verdite, 322
Verneuil, A., 207
Visible light, 74–76
Vitreous lustre, 87
Volcanic rock, 10, 12, 170, 194
Vulcanite, 307

Walrus ivory, 309–310
Water opal, 197
Watermelon tourmaline, 186
Wave theory, 73–77
Wavelength, 74–77
Waxing, 199, 253–254
Waxy lustre, 87–88
Wear (signs of hardness), 50, 164
White opal, 197–198
Wiedemann, E., 142

X-ray diffraction, 155, 296
X-ray fluorescence, 143, 156
X-ray transparency, 155–156
X-rays, 75, 154–157

YAG, 216, 222
Yehuda, 262–264, 265
Yttrium aluminium garnet (YAG), 216, 222

Zircon, 106, 189–190, 254–257
    Spectrum, 138, 190
Zircon cut, 280
Zoisite, 322–323